Multisector Growth Models

Terry L. Roe · Rodney B.W. Smith ·
D. Şirin Saracoğlu

Multisector Growth Models

Theory and Application

Springer

Terry L. Roe
Department of Applied Economics
University of Minnesota
1994 Buford Ave.
St. Paul MN 55108
USA
troe@tc.umn.edu

Rodney B.W. Smith
Department of Applied Economics
University of Minnesota
1994 Buford Ave.
St. Paul MN 55108
USA
smith142@umn.edu

D. Şirin Saracoğlu
Department of Economics
Middle East Technical University (METU)
Inonu Bulv.
06531 Ankara
Turkey
ssirin@metu.edu.tr

ISBN 978-0-387-77357-5 e-ISBN 978-0-387-77358-2
DOI 10.1007/978-0-387-77358-2
Springer New York Dordrecht Heidelberg London

Library of Congress Control Number: 2009935051

Printed on acid-free paper

Springer is part of Springer Science+Business Media (www.springer.com)

Contents

Preface

The primary objective of this book is to advance the state of the art in specifying and fitting to data structural multi-sector dynamic macroeconomic models, and empirically implementing them. The fundamental construct upon which we build is the Ramsey model. A most attractive feature of this model is the insights it provides into the dynamics of an economy in transition to long-run equilibrium. With some exceptions, Ramsey models are highly aggregated – typically single sector models. However, interest often lies in understanding the forces of economic growth across multiple sectors of an economy and on how policy impacts likely play out over time. Such analyses call for more disaggregated models that can be fit to country or regional data. This book shows how to: (i) extend the basic model to multiple sectors, (ii) how to adapt the basic model to account for policy instruments, and (iii) fit the model to data, and obtain equilibrium values both forward and backward in time from the data points to which the model is initially fit.

Although extremely helpful in understanding economic growth and structure, theory alone is not sufficient; we also need to confront theory with data. Fitting growth models to data has been greatly facilitated by advances in numerical algorithms and computer technology. The ease of obtaining numerical solutions using procedures that are relatively robust across a broad range of model specifications is important because the differential equations of even the single sector, two factor, closed economy Ramsey model are essentially analytically intractable. The methods advanced here, and illustrated with numerical examples, are easily used in the classroom. Our experience suggests this material is accessible to advanced undergraduate and beginning graduate students, and easily managed by those working in agencies and bureaus familiar with general equilibrium concepts. An understanding of the subtleties of control theory and numerical algorithms is not required, but familiarity with a programming language such as Mathematica is essential. Over the past several years, we have had students choose a country, conduct a growth accounting exercise, formulate the country data into an elementary social accounting matrix format, fit a model to data, and then obtain a numerical solution using off-the-shelf software. We found that such assignments greatly strengthens students' grasp of theoretical concepts and helps them link these concepts to real economies. Grasping the theory and knowing how to implement the theory to obtain empirical insights into real problems provides them a form of human capital that they are unlikely to attain so easily in other ways.

The book is organized by first reviewing the fundamentals of duality theory of the consumer and firm, which is then used to review the standard two-sector, two-factor Heckscher-Ohlin-Samuelson model of a small open economy. Using duality theory, Chapter 3 introduces the two sector closed economy Ramsey model in a rather structured fashion, and concludes with an empirical example. Chapter 4 develops a three-sector, open economy model with a non-traded good sector. Chapters 5 and 6 extend the three-sector model in several directions: intermediate factors of production; capital stock composed of the output of all sectors of the economy; government consumption with taxes and lump-sum transfers from households. Chapter 7 concludes with a two country "world" model. In each chapter, the model presentation follows a similar pattern and builds off the structure of the previous chapter. Each modeling chapter concludes with an empirical example using the same data set. The book concludes with two chapters that discuss how the data are organized to facilitate the fitting of models to data, and the strategy used to facilitate the solution of each model's system of differential equations.

Acknowledgement

We benefited from over 50 former graduate students who contributed their talent, insights and wit in workshops, seminars and reviews of earlier drafts of this book. Special mention goes to Xinshen Diao, Elamin Elbasha, Beatriz Gaitan, Xavier Irz, and Erinc Yeldan. We thank Harumi Nelson for her special contribution. We benefited from many intellectual exchanges with Professors V. V. Chari, Timothy Kehoe, and Anne Krueger. We also express appreciation to Bruce Gardner, James Roumasset, Mathew Shane, and Agapi Somwaru.

Personal acknowledgements:

Terry L. Roe to Carol, Mark and Seth

Rodney Smith to Margaret Smith, D.T. Luck and T.S. Khanna

Şirin Saracoğlu to Family

1

Introduction: Orientation and Focus

1.1 Introduction

Our general objective is to advance methodology for the analyses of economies using multi-sector dynamic general equilibrium models. We begin by building upon the static Heckscher-Ohlin-Samuelson framework expressed using duality theory as shown most clearly by Woodland (1982). This framework is then cast into a two-sector growth model in which infinitely lived households choose consumption and saving to maximize their dynastic utility. This specification of consumer behavior is a key element of the Ramsey growth model (as constructed by Ramsey (1928) and refined by Cass (1965) and Koopmans (1965)), and has now become the mainstay of growth models with endogenous savings behavior. We then extend this two-sector framework to less stylistic models over several chapters that culminate in a two country world model where each country produces two traded goods and a home-good.

These models are specified in continuous time and in a manner that proceeds from the specification of primitives, the definition of equilibrium, and the characterization of intra-temporal and inter-temporal equilibrium. This manner of presentation is desirable for several reasons. First, it greatly facilitates an understanding of the analytical features of each model, including "Stopler-Samuelson-like" and "Rybczynski-like" relationships similar to that of the static trade model. Second, it shows more clearly how a less stylized model follows from the former; thus providing insights into how further extensions of the framework not discussed here or envisioned by the authors might be developed. Third, this style of presentation facilitates the writing of code to empirically solve a model because it lays out clearly:

T.L. Roe et al., *Multisector Growth Models*, DOI 10.1007/978-0-387-77358-2_1,

(i) the key relationships among equations characterizing equilibrium that must be coded, (ii) the parameters to be estimated, and (iii) the necessary initial and terminal conditions for equilibrium. Moreover, since a challenge of writing code is one of finding coding errors, this process allows for virtually each line of code to be checked to determine whether it is reproducing the data corresponding to one or more of the model's structural equations.

Each chapter illustrates an application of the theory with an empirical example. The empirical examples draw upon a common data set, organized into a logically consistent structure and at a level of aggregation equivalent to each analytical model. This structure, which is well known to those familiar with static computable general equilibrium (CGE) analyses, is laid out in Chapter 8 and shows how to estimate many of the model's parameters in such a way that the empirical results approximate the base period values of an economy in transition to long-run growth. The empirical models can be solved both backward and forward in time, and thus the question naturally arises: Do model results replicate the data over some time interval? We thus discuss the need to validate model results with time series data, identify a number of challenges that remain in this regard, and contrast results from one of the less stylistic models to the underlying data.

Static CGE analysis was made possible by advancements in solution techniques like Scarf's fixed point algorithm, and later, software to solve collections of non-linear algebraic equations. Another facilitating factor was the development of the social accounting matrix as a way to reconcile the national income and product accounts with input-output accounts.[1] These developments, along with the more recent development of the Global Trade, Assistance, and Production (GTAP) data base,[2] among

[1]See Robinson (1989) for a review of literature of static computable general equilibrium models and a discussion of the structure of a social accounting matrix.

[2]This data base is supported by the Center for Global Trade Analysis, Dept. of Agricultural Economics, Purdue University with funding from numerous international agencies and government sources. The earlier versions of the data base built heavily

others, have made possible a plethora of empirical general equilibrium models. The proliferation of these models has been accompanied by critiques that the models lack empirical realism and are based on ad hoc algebraic structures with little or no supporting conceptual framework. For these and other reasons, Kehoe (2003) suggests ex-post performance evaluations of such models are essential if policy makers are to have confidence in the results produced by them.

We see a potential parallel with this history and the next generation of dynamic models as a challenge faced by this book. Two sets of developments help us face this challenge. One is the recent advances in computer software such as Mathematica, Matlab, GAUSS and others, each having built-in subroutines for solving systems of differential equations. The other is the development of solution strategies like the time elimination method of Mulligan and Sala-i-Martin (1991, 1993), the relaxation algorithm implemented by Trimborn et al. (2008), and the backward integration method of Brunner and Strulik (2002).[3] These developments suggest the feasibility of moving beyond the static models, and the ability to not repeat past mistakes. Part of this problem is rectified by the requirement that the empirical model to draw directly upon the inter-temporal equilibrium conditions of the analytical model, thus avoiding some of the criticisms related to ad hoc model specifications. Another rectifying effect that is not as prominent in static models is the natural tendency to contrast a dynamic model's predictions with time series data, thus pressuring the development of validation procedures.

It is generally accepted that the accumulation of physical and human capital are important to the economic growth of economies, but these effects only explain part of the variation across countries in income per capita and differences in rates of growth. Including exogenous total factor productivity, as we do in the models developed in this book, greatly improves the

---

on the SALTER Project which was undertaken at the Australian Industry Commission during the 1980s and early 1990s. See Badri and Walmsley (2008).

[3]The Handbook of Computational Economics, volume 2, edited by Tesfatsion and Judd (2006), provides a thorough discussion on the solution of saddlepoint problems.

model's fit to data, but this measure remains an "ignorance" type parameter. Roughly, economic growth models of recent vintage, e.g., Romer (1990), and those presented by Aghion and Howitt (1998, 2009), question factor accumulation as the main engine of growth, focusing instead on the notion that growth is primarily driven by innovations that are themselves the result of profit motivated activities. Nevertheless, these models tend to remain highly stylized, and not particularly useful for the analysis of foreign trade and other policies of common importance to policy makers. We suggest that the topics and procedures addressed in this book are important precursors to the future empirical application of endogenous growth models.

1.2 Organization of the book

The chapters are laid out as follows. Chapter 2 focuses on static microeconomic theory and trade theory of the small open economy. The first part sets out the microeconomic principles and conditions used throughout the book with emphasis on duality theory of the consumer and the firm. These microeconomic relationships are then aggregated to the economy level where we specify the Heckscher-Ohlin-Samuelson (HOS) two sector, two factor, small open economy model. The Stopler-Samuelson and Rybczynski theorems are shown and the point made that the characterization of equilibrium of this model, and the theorems play a role in helping us specify equilibrium conditions and interpret the transition dynamics of models developed in later chapters. The chapter concludes with a static model of a three sector, small and open economy that produces a non traded good. This model establishes the endogenous price of the home-good as a function of the prices of traded goods and factor endowments. These and all other models are presented using duality theory.

Chapter 3 introduces the two-sector Ramsey model of a closed economy in a series of steps. Each step helps acquaint the reader with: (i) the specification of household and firm optimization,

(ii) the normalization of variables into effective labor units, and (iii) the characterization of intra-temporal and inter-temporal equilibrium using duality in much the same manner as the HOS model. This pattern of model development is followed in all chapters. The chapter concludes with an algebraic example, and an empirical example. All of the empirical examples from Chapters 3, 4, 5, 6 and 7 are purposefully designed to link the empirical example of one to that of the other. Mathematica is used to obtain numerical solutions in all cases. The examples draw upon the same data of the Turkish economy for the year 2001, with the level of aggregation chosen to fit the structure of the analytical model in each chapter. The data and numerical procedures to support and illustrate how the models are fit to data and solved are presented in the last two chapters.

Chapter 4 presents the three sector model with two traded goods, one non-traded good and three factors of production. The intra-temporal equilibrium conditions for this model closely parallel the conditions of the static three sector model of Chapter 2, so a number of comparative static properties of the static model carry over to the dynamic model. The numerical example suggests that about 27 years are required for income per Turkish worker to double, Rybczynski-like effects of transition growth are shown as is the decline in the share of labor in agriculture and a number of other features of the Turkish economy.

Chapter 5 considers a number of extensions to the basic three sector model of Chapter 4, each of which are considered separately. These include intermediate inputs, vertical market structures, composite capital, and government revenues and expenditures. The composite capital specification is motivated by data which show that a country's stock of capital is a composite of the outputs of almost all sectors of the economy. The empirical example focuses on the intermediate input extension to the model where changes in the internal terms of trade during transition growth are shown to affect the traded goods sectors of the economy. Then, since the home-good is an important intermediate input to all sectors, a simulated productivity shock to this sector is shown to have multiplier effects on the traded good sectors that can be interpreted as releasing resources over time from home-good production to increase the country's gains from trade.

Chapter 6 incorporates all of the extensions of Chapter 4 into a single model. A number of complexities not encountered in Chapter 4 are encountered and addressed. The chapter also discusses a number of issues related to validating the empirical model. The model is solved both backward and forward from the base period 2001 and the results contrasted with time series data on Turkish gross domestic output and sectoral value-added for the period 1995–2006. A simulation is also performed which shows the effect on transition growth from lowering the tariff rate protecting the industrial sector.

Chapter 7 draws upon the model of Chapter 4 and casts it into a two-country-world in which each country produces two traded goods and their own respective non-traded good. Two models are considered. In the first, residents of one country can hold claims to capital stock in the other country over time. The second model disallows foreign ownership of domestic assets, in which case the model has two state variables, the capital stock in each country. To maintain a sense of symmetry with the previous numerical examples, we show how the capital stock of one country can be expressed as a function of the stock of the other country. This observation reduces our system to a single state variable. However, the number of differential equations characterizing inter-temporal equilibrium are increased to four, which nevertheless, can be numerically solved using the time elimination method. This chapter also concludes with a numerical example which focuses on the model permitting capital flows between countries.

Chapter 8 links the data to the model. We show how the Turkish data are organized into a social accounting matrix and how the parameters of the various empirical examples are estimated. The chapter also provides a definition of the economy's economic sectors.

Chapter 9 addresses the question of how to use numerical methods to solve the empirical models. The time elimination method receives considerable attention since this method is used in all of the empirical examples. We show, using Mathematica, a few coding "tricks" to facilitate solution and mention a number of other methods that can also be used to solve these models.

2

The Preliminaries

This chapter discusses features of static trade theory that are important components of the dynamic, multi-sector models developed in later chapters. Most of the notation used throughout the text is introduced and the style used to state the model's primitives, and to define and characterize equilibrium is presented.

The first section reviews key concepts and results from individual consumer and producer theory relevant to neoclassical trade theory. The exposition is simplified by assuming production technologies and preferences are differentiable and homothetic functions. Throughout the text we draw heavily upon the so called dual or indirect functions that characterize the constrained optimization behavior of individual agents. Readers interested in a more rigorous exposition of consumer theory should consult Cornes (1992) or Mas-Colell et al. (1995). A more rigorous treatment of producer theory can be found in Chambers (1988), and Fare and Grosskop (2004).

Using the concepts developed in Sections 1 and 2 introduces the Heckscher-Ohlin-Samuelson (HOS) model of a small open and competitive economy. The basic features of equilibrium and comparative statics as provided by the Stopler-Samuelson and Rybczynski theorems are discussed. Woodland (1982) provides an excellent characterization of this model. Section 3 considers, briefly, some further generalizations of the comparative statics of the HOS model. Section 4 concludes this chapter and presents a model of two traded goods, a home-good and three factors of production. A dynamic version of this model follows in later chapters.

T.L. Roe et al., *Multisector Growth Models*, DOI 10.1007/978-0-387-77358-2_2,

2.1 Microeconomic foundations

Throughout the text, the following notation denotes factor endowments, factor rental rates and output prices. Sectors are indexed by $j \in \{1, ..., M\}$, and denote the quantity of sector-j's output by the scalar Y_j. Corresponding output prices are denoted $\mathbf{p} = (p_1, ..., p_M) \in \mathbb{R}^M_{++}$, with the scalar p_j representing the per-unit price of sector-j output. We index factor endowments by $i \in \{1, ..., N\}$, and denote the economy's level of endowment i by the scalar V_i and the vector of factor endowments by $\mathbf{V} \equiv (V_1, ..., V_N) \in \mathbb{R}^N_{++}$. Corresponding factor rental rates are denoted $\mathbf{w} = (w_1, ..., w_N) \in \mathbb{R}^N_{++}$, with the scalar w_i representing the rental rate of factor V_i. For simplicity, outputs are often given a sector specific designation, such as agriculture, a, manufacturing, m, and the home-good, s. Likewise, endowments are often given designations like labor, L, capital, K, and land H.

2.1.1 *Consumer preferences*

The economy is composed of a large number of atomistic households. Each household faces the same vector of prices $\mathbf{p}$ and the same vector of factor rental rates $\mathbf{w}$. Let $\boldsymbol{v}^h = \left(v_1^h, ..., v_N^h\right) \in \mathbb{R}^N_{++}$ denote the level of factor endowments held by household-h, with v_i^h representing the household's endowment of factor i. In most applications that follow we suppress the h superscript of $\boldsymbol{v}^h$ and v^h, and use instead $\boldsymbol{v}$ and v_i. Given factor rental rates $\mathbf{w}$, the household's income is given by $\mathbf{w} \cdot \boldsymbol{v}$, which is used to purchase q_j units of consumption good j at market price p_j, $j = 1, ..., M$. Then, the household's budget constraint is given by

$$\mathbf{w} \cdot \boldsymbol{v} \geq \mathbf{p} \cdot \mathbf{q}$$

where $\mathbf{q} = (q_1, ..., q_M) \in \mathbb{R}^M_{++}$. In other words, each household consumes a strictly positive level of each consumption good.

Consumer preferences over goods are represented by the utility function $u : \mathbb{R}^M_{++} \to \mathbb{R}_+$, defined as $u(\mathbf{q})$.

Assumption 1 $u(\mathbf{q})$ *satisfies the following properties:*

1. $u(\mathbf{q})$ is increasing and strictly concave in $\mathbf{q}$,
2. $u(\mathbf{q})$ is everywhere continuous, and everywhere twice differentiable,
3. $u(\mathbf{q})$ is homothetic.

Assumption 1.1 yields indifference curves that are convex, Assumption 1.2 ensures Marshallian demands are continuous functions, while Assumption 1.3 yields Marshallian demands that are separable in prices and income.

Two indirect functions emerge from the consumer's problem: the indirect utility function and the expenditure function. The *indirect utility function* gives the household's maximum attainable utility given income $\mathbf{w}\cdot\boldsymbol{\upsilon}$, defined as

$$\mathcal{V}(\mathbf{p}, \mathbf{w}\cdot\boldsymbol{\upsilon}) \equiv \max_{\mathbf{q}} \{u(\mathbf{q}) : \mathbf{w}\cdot\boldsymbol{\upsilon} \geq \mathbf{p}\cdot\mathbf{q}\}$$

The indirect utility function inherits the following properties from the direct utility function (see Cornes, pp. 67–70):

V1. Homogeneous of degree zero in $\mathbf{p}$ and $\mathbf{w}\cdot\boldsymbol{\upsilon}$; $\mathcal{V}(\theta\mathbf{p}, \theta\mathbf{w}\cdot\boldsymbol{\upsilon}) = \mathcal{V}(\mathbf{p}, \mathbf{w}\cdot\boldsymbol{\upsilon})$, $\theta > 0$,

V2. $\mathcal{V}(\mathbf{p}, \mathbf{w}\cdot\boldsymbol{\upsilon})$ is convex in $\mathbf{p}$,

V3. $\mathcal{V}(\mathbf{p}, \mathbf{w}\cdot\boldsymbol{\upsilon})$ is continuous and differentiable in $\mathbf{p}$ and $\mathbf{w}\cdot\boldsymbol{\upsilon}$,

V4. $\mathcal{V}(\mathbf{p}, \mathbf{w}\cdot\boldsymbol{\upsilon}) = v(\mathbf{p})\,\mathbf{w}\cdot\boldsymbol{\upsilon}$: separable in $\mathbf{p}$ and $\mathbf{w}\cdot\boldsymbol{\upsilon}$,

By V4, the marginal utility of an additional unit of income is $v(\mathbf{p})$.

V5. Given differentiability, Marshallian demands follow from Roy's identity,

$$q^j(\mathbf{p})(\mathbf{w}\cdot\boldsymbol{\upsilon}) = -\frac{v_{p_j}(\mathbf{p})}{v(\mathbf{p})}\mathbf{w}\cdot\boldsymbol{\upsilon} \tag{2.1}$$

where, throughout the text, the subscript on a function indicates a partial derivative, e.g., $v_{p_j} = \partial v(\mathbf{p})/\partial p_j$ and $v_{p_1 p_2} = \partial^2 v(\mathbf{p})/\partial p_1 \partial p_2$.

Since consumers face the same prices and have identical preferences, the "community" indirect utility function is given by

$$\mathcal{V} = v(\mathbf{p})(\mathbf{w} \cdot \mathbf{V})$$

while the total domestic Marshallian demand for good j is

$$Q_j = q^j(\mathbf{p})(\mathbf{w} \cdot \mathbf{V}), \ \forall j \in \{1, \cdots, M\} \tag{2.2}$$

These functions are the simple aggregation of individual consumer welfare and demands. It also follows from V1 that (2.2) is homogeneous of degree minus one in prices $\mathbf{p}$ and of degree one in income.

The *expenditure function* gives the minimum cost of achieving utility level $q \in \mathbb{R}$ at given prices $\mathbf{p}$, and is defined as

$$E(\mathbf{p}, q) \equiv \min_{\mathbf{q}} \{\mathbf{p} \cdot \mathbf{q} : q \leq u(\mathbf{q})\}$$

The expenditure function inherits from $u(\cdot)$, the following properties:

E1. $E(\mathbf{p}, q) > 0$ for any $\mathbf{p}$ and $q > 0$,

E2. $E(\mathbf{p}, q)$ is non-decreasing in $\mathbf{p}$ and q,

E3. $E(\mathbf{p}, q)$ is concave and continuous in $\mathbf{p}$,

E4. $E(\lambda\mathbf{p}, q) = \lambda E(\mathbf{p}, q)$, $\lambda > 0$: homogeneous of degree 1 in $\mathbf{p}$,

E5. $E(\mathbf{p}, q) = \mathcal{E}(\mathbf{p}) q$: separable in $\mathbf{p}$ and q,

E6. Shephard's lemma: If $E(\mathbf{p}, q)$ is differentiable in $\mathbf{p}$, then

$$q_j = E_{p_j}(\mathbf{p}, q) = \mathcal{E}_{p_j}(\mathbf{p}) q, \ j = 1, ..., M$$

E1 says purchasing a strictly positive consumption bundle is costly. **E**2 says, all else equal, (i) if the price of a consumption good increases, then the cost of achieving the same level of utility increases, or (ii) increasing utility requires an increase in expenditures. By **E3**, the expenditure function is continuous and yields downward sloping Hicksian demand functions. Condition **E4** implies demand functions are homogeneous of degree zero in $\mathbf{p}$. **E5** results from Assumption 1.3 and implies demand functions are separable in $\mathbf{p}$ and q (see Chambers, 1988, Chapter 2). Later, we interpret the quantity q to be a composite consumption good, the unit cost of which is $\mathcal{E}(\mathbf{p})$.

2.1.2 Production technologies

Each sector j is composed of a large number of identical, atomistic firms. Each firm faces the same vector of input and output prices. Let y_j be the output of each firm in sector j and let $\boldsymbol{v}^j \equiv \left(v_1^j, ..., v_N^j\right) \in \mathbb{R}_+^N$ represent the vector of productive factors used by that firm, where v_i^j is the level of factor i used by the sector j firm. Represent the technology of a sector j firm by the production function $f^j : \mathbb{R}_+^N \rightarrow \mathbb{R}_+$, defined as $y_j = f^j\left(\boldsymbol{v}^j\right)$.

Assumption 2 *$f^j\left(\boldsymbol{v}^j\right)$ satisfies the following properties:*

1. $f^j(\mathbf{0}) = 0$, and $f^j\left(\boldsymbol{v}^j\right) > 0$ for any $\boldsymbol{v}^j \gg \mathbf{0}^N$,
2. $f^j\left(\boldsymbol{v}^j\right)$ is linearly homothetic in $\boldsymbol{v}^j$,
3. $f^j\left(\boldsymbol{v}^j\right)$ is non-decreasing and strictly concave in $\boldsymbol{v}^j$,
4. $f^j\left(\boldsymbol{v}^j\right)$ is everywhere continuous and everywhere twice differentiable in $\boldsymbol{v}^j$.

Here $\mathbf{0}^N \in \mathbb{R}_+^N$ is a vector of N zeros and the notation $\boldsymbol{v}^j \gg \mathbf{0}$ means at least one element of $\boldsymbol{v}^j$ is strictly positive. Assumption 2.1 ensures it is not possible to produce a positive level of output with no input, and ensures there are no fixed costs. Assumption 2.2 says individual firm technologies satisfy constant returns to scale (CRS). An important implication of Assumption 2.2 is, when all firms face the same output and input prices, sectoral production levels and input demands are simple linear aggregations of individual firm choices. Another implication is the corresponding cost function is separable in input prices and output levels. Assumption 2.3 ensures the production technology is well-behaved and yields the familiar convex isoquants: it imposes diminishing marginal returns on individual input use. Assumptions 2.1 and 2.3 also ensure the existence of a cost and aggregate value-added (GDP) function defined later. Finally, Assumption 2.4 allows the use of differential calculus to derive corresponding cost and GDP functions.

Two indirect functions associated with the producer's problem are the cost function and the sectoral value-added function. The *cost function* is defined as:

$$c^j(\mathbf{w}, y_j) \equiv \min_{\boldsymbol{v}^j} \left\{ \mathbf{w} \cdot \boldsymbol{v}^j : y_j \leq f^j\left(\boldsymbol{v}^j\right) \right\}, \ j = 1, 2, ..., M$$

and is the firm's analog of the household's expenditure function. It inherits from Assumption 2, the following properties:

C1. $c^j(\mathbf{w}, y_j) > 0$ for any $\mathbf{w}$ and $y_j > 0$,

C2. $c^j(\mathbf{w}, y_j)$ is non-decreasing in $\mathbf{w}$ and y_j,

C3. $c^j(\mathbf{w}, y_j)$ is concave and continuous in $\mathbf{w}$,

C4. $c^j(\theta\mathbf{w}, y_j) = \theta c^j(\mathbf{w}, y_j)$: homogeneous of degree one in $\mathbf{w}$,

C5. $c^j(\mathbf{w}, y_j) = C^j(\mathbf{w})\, y_j$: separable in $\mathbf{w}$ and y_j,

where $C^j(\mathbf{w})$ is the unit cost of producing output j. Finally, we have

C6. Shephard's lemma: If $c^j(\mathbf{w}, y_j)$ is differentiable in $\mathbf{w}$, then

$$v_i^j = C_{w_i}^j(\mathbf{w})\, y_j, \ i = 1, ..., N,$$

where $C_{w_i}^j(\cdot)$ is the derived unit demand for input i from sector j.

C1 says producing a strictly positive level of output is costly. **C2** says, all else equal, if the price of an input increases production cost increases, or increasing output increases production costs. By **C3**, the cost function is continuous and yields conditional input demand functions that are decreasing in own prices. Condition **C4** implies input demand functions are homogeneous of degree zero in $\mathbf{w}$. **C5** results from Assumption 2.2 and implies constant marginal and average costs. Furthermore, given **C5**, the output supply and input demand functions are both separable in $\mathbf{w}$ and y_j (see Chambers, 1988, Chapter 2).

Since all firms in a sector employ the same technology and face the same output and input prices, characterizing the aggregate technology for the sector is straightforward. Let $\mathbf{V}^j \equiv (V_1^j, ..., V_N^j) \in \mathbb{R}_+^N$ denote the vector of factors employed in producing output Y_j, where V_i^j is the aggregate level of factor i used by sector j firms. While the total number of firms in a sector are indeterminate, their identical nature implies if each firm produces a share, Υ_j^o, of total sectoral output j, then the firm also employs the same $\bar{\Upsilon}_j$ share of factor inputs, i.e.,

$$\begin{aligned} y_j^o &= \Upsilon_j^o Y_j, \\ v_i^o &= \Upsilon_j^o V_i^j, \ \forall i \in \{1, \cdots, N\} \end{aligned}$$

Hence, the sector level production function is a linear expansion of individual firm production functions. That is,

$$\Upsilon_j^o Y_j = f^j\left(\boldsymbol{v}^j\right) = f^j\left(\Upsilon_j^o \mathbf{V}^j\right)$$

which implies the sector level production function is

$$Y_j = f^j\left(\mathbf{V}^j\right)$$

To distinguish between firm level and aggregate sectoral production however, it is convenient to represent the aggregate technology for sector j by the production function $\mathcal{F}^j : \mathbb{R}_+^N \to \mathbb{R}_+$, defined as

$$Y_j = \mathcal{F}^j\left(\mathbf{V}^j\right) \tag{2.3}$$

Then, the corresponding sectoral cost function, denoted TC_j, is given by

$$TC_j = C^j\left(\mathbf{w}\right) Y_j \tag{2.4}$$

The economy-wide gross national product function is obtained by maximizing aggregate sectoral income subject to the technology (2.3) and the endowment constraints. In this case we have:

$$G(\mathbf{p},\mathbf{V}) \equiv \max_{\mathbf{V}^1,\ldots,\mathbf{V}^M} \left\{ \sum_{j=1}^{M} p_j \mathcal{F}^j\left(\mathbf{V}^j\right) : V_i \geq \sum_{j=1}^{M} V_i^j,\ i = 1,\ldots,M \right\} \tag{2.5}$$

Woodland (1982, p. 123) shows the function $G(\cdot)$ satisfies the following properties:

G1. $G(\mathbf{p},\mathbf{V}) \geq 0$ for all $\mathbf{p}$ and $\mathbf{V}$,

G2. $G(\lambda\mathbf{p},\mathbf{V}) = \lambda G(\mathbf{p},\mathbf{V})$, $\lambda > 0$: linearly homogeneous in $\mathbf{p}$,

G3. $G(\mathbf{p},\lambda\mathbf{V}) = \lambda G(\mathbf{p},\mathbf{V})$ $\lambda > 0$: linearly homogeneous in $\mathbf{V}$,

G4. $G(\mathbf{p},\mathbf{V})$ is continuous, non-decreasing, and convex in $\mathbf{p}$,

G5. $G(\mathbf{p},\mathbf{V})$ is continuous, non-decreasing, and concave in $\mathbf{V}$,

G6. Hotelling's lemma. If $G(\cdot)$ is everywhere differentiable in $\mathbf{p}$ and $\mathbf{V}$, then

$$\begin{aligned} Y_j &= G_{p_j}(\mathbf{p},\mathbf{V}) \\ w_i &= G_{V_i}(\mathbf{p},\mathbf{V}) \end{aligned}$$

The major implications of conditions G1 – G6 are that the gradients of $G(\cdot)$ yield aggregate sectoral supply functions, $G_{p_j}(\mathbf{p},\mathbf{V})$, that are non-decreasing in own-price, homogeneous of degree zero in prices $\mathbf{p}$, and homogeneous of degree one in endowments $\mathbf{V}$. The inverse factor demand functions $G_{V_i}(\mathbf{p},\mathbf{V})$ are downward sloping in own factor levels, homogeneous of degree one in prices and homogeneous of degree zero in endowments. The Hessian matrix of $G(\mathbf{p},\mathbf{V})$ is positive semi-definite.[1]

[1] Young's theorem implies that the second derivative matrix of $G(\mathbf{p},\mathbf{V})$ is symmetric, $G_{\mathbf{pV}}(\cdot) = G_{\mathbf{Vp}}(\cdot)$. Thus, an increase in w_i due to a unit increase in p_j is equal to the increase in Y_j due to an increase in v_i. See Diewert (1973, 1974).

It is also convenient to specify a sectoral value-added function. For many of the models developed in this text, at least one productive sector is endowed with a factor specific to its production process. For example, we typically model land as a factor used only in producing agricultural products. Farmers can rent land in and out among themselves at some market determined land rental rate, but they do not rent land to producers in other sectors of the economy. Since each farmer's production function satisfies Assumption 2, there exists a corresponding sectoral agricultural production and cost function (2.3) and (2.4). However, in the case of the sectoral production function, the sector specific factor is pre-determined or fixed, and the sectoral level production function exhibits decreasing returns to scale in all the other factors employed in other sectors of the economy. This property gives rise to a sector level value-added function.

More formally, divide the input vector $\mathbf{V}^j$ into two subvectors, a vector of variable inputs and a vector of sector specific inputs. Let the first ς_j factors be variable and the remaining $N - \varsigma_j$ factors be sector specific. Denote the vector of variable factors by $\mathbf{V}^j = \left(V_1^j, ..., V_{\varsigma_j}^j\right) \in \mathbb{R}_+^{\zeta_j}$, and denote the vector of sector specific factors by $\bar{\mathbf{V}}^j = \left(\bar{V}_{\zeta_j+1}^j, ..., \bar{V}_N^j\right) \in \mathbb{R}_+^{N-\zeta_j}$. With fixed factors $\bar{\mathbf{V}}^j$, the jth sector value-added function can be defined as:

$$\Pi^j \left(p_j, \mathbf{w}, \bar{\mathbf{V}}^j\right) \equiv$$

$$\max_{y_j, \mathbf{V}^j} \left\{p_j Y_j - \mathbf{w} \cdot \left(\mathbf{V}^j, \mathbf{0}^{N-\varsigma_j}\right) : Y_j \geq \mathcal{F}^j \left(\mathbf{V}_\varsigma^j, \bar{\mathbf{V}}^j\right)\right\} \tag{2.6}$$

where $\mathbf{0}^{N-\varsigma_j} \in \mathbb{R}_+^{N-\zeta_j}$ is a vector of zeros. Given Assumption 2, the sectoral value-added function properties include:

Π1. $\Pi^j \left(p_j, \mathbf{w}, \bar{\mathbf{V}}^j\right) \geq 0$ for all $p_j, \mathbf{w}$, and $\bar{\mathbf{V}}^j$,

Π2. $\Pi^j \left(p_j, \mathbf{w}, \bar{\mathbf{V}}^j\right)$ is nondecreasing in p_j and nonincreasing in $\mathbf{w}$,

Π3. $\lambda \Pi^j \left(p_j, \mathbf{w}, \bar{\mathbf{V}}^j\right) = \Pi^j \left(\lambda p_j, \lambda \mathbf{w}, \bar{\mathbf{V}}^j\right)$, $\lambda > 0$: linearly homogeneous in p_j and $\mathbf{w}$,

Π4. $\lambda\Pi^j\left(p_j, \mathbf{w}, \bar{\mathbf{V}}^j\right) = \Pi^j\left(p_j, \mathbf{w}, \lambda\bar{\mathbf{V}}^j\right), \lambda > 0$: linearly homogeneous in $\bar{\mathbf{V}}^j$

Π5. $\Pi^j\left(p_j, \mathbf{w}, \bar{\mathbf{V}}^j\right) = \pi^j\left(p_j, \mathbf{w}\right)\Phi\left(\bar{\mathbf{V}}^j\right)$: separable in fixed endowments,

Π6. Hotelling's lemma. If $\Pi^j(\cdot)$ is everywhere differentiable in $\mathbf{p}$, $\mathbf{w}$ and $\bar{\mathbf{V}}^j$, then sectoral supply Y_j and sectoral factor demand V_i^j are, respectively,

$$\begin{aligned} Y_j &= \Pi^j_{p_j}\left(p_j, \mathbf{w}, \bar{\mathbf{V}}^j\right) \\ V_i^j &= -\Pi^j_{\bar{w}_i}\left(p_j, \mathbf{w}, \bar{\mathbf{V}}^j\right) \end{aligned}$$

The factor rental rate (or shadow price) of the sector specific factors is given by

$$w_i^j = \Pi^j_{\bar{V}_i^j}\left(p_j, \mathbf{w}, \bar{\mathbf{V}}^j\right)$$

For the case of a single sector specific factor, say land, that is rented in or out among farmers, $\pi^j\left(p_j, \mathbf{w}\right)$ is the rental rate that clears the land rental market. Moreover, it can be shown that the output price gradient of the economy-wide GDP function yields the same level of supply as the corresponding output price gradient of the sector value-added function,

$$Y_j = G_{p_j}\left(\mathbf{p}, \mathbf{V}\right) = \Pi^j_{p_j}\left(p_j, \mathbf{w}, \bar{\mathbf{V}}^j\right)$$

where the gradients are evaluated at values $(\mathbf{p}, \mathbf{w}, \mathbf{V})$ yielding an equilibrium to the economy. This property is particularly useful for decomposing effects into direct and indirect. For instance, the direct effect of a price change on Y_j is $\partial\Pi^j_{p_j}/\partial p_j$ while the indirect effects are transmitted through factor markets and are given by $\left(\partial\Pi^j_{p_j}/\partial w_i\right)\left(\partial w_i/\partial p_j\right)$. Together, they equal the total effect which can be shown to equal $\partial G_{p_j}\left(\mathbf{p}, \mathbf{V}\right)/\partial p_j$.

2.2 The Heckscher-Ohlin-Samuelson model

The optimizing behavior of producers and consumers embodied in expressions (2.2) and (2.4) provide the building blocks for specifying the well known Heckscher-Ohlin-Samuelson (HOS) model. The economy is small, open and competitive, endowed with two factors, and produces two outputs. Denote the endowment vector by $\mathbf{V} = (L, K)$, and interpret K as units of physical capital and interpret L as units of labor. Neither endowment is traded internationally. A main feature of the model is that the number of traded goods equal the number of factors, $M = N$.

2.2.1 The behavior of households

The individual household[2] is endowed with resources $\boldsymbol{v} = (\ell, k) \in \mathbb{R}^2_{++}$, where ℓ and k denote labor and capital, respectively. The household provides the services of these resources to firms in return for wages, w, and capital rents, r, yielding income $w\ell + rk$.

Given prices (p_1, p_2), the household's budget constraint is

$$w\ell + rk \geq p_1 q_1 + p_2 q_2$$

Consumer preferences are given by the utility function $u(q_1, q_2)$ satisfying Assumption 1. Consequently, the consumer's optimization problem yields the indirect utility function

$$\mathrm{v}(p_1, p_2)(w\ell + rk) \equiv \max_{q_1, q_2} \{u(q_1, q_2) : w\ell + rk \geq p_1 q_1 + p_2 q_2\}$$

where $\mathrm{v}(p_1, p_2)(w\ell + rk)$ satisfies properties V1 – V4. The corresponding Marshallian demands are ,

$$q^j(p_1, p_2)(w\ell + rk) = -\frac{\mathrm{v}_{p_j}(p_1, p_2)}{\mathrm{v}(p_1, p_2)}(w\ell + rk), \; j = 1, 2$$

Since consumers face the same prices and have identical preferences, the community indirect utility function is

$$\mathcal{V} = \mathrm{v}(p_1, p_2)(wL + rK)$$

while aggregate domestic Marshallian demand for good j is

$$Q_j = q^j(p_1, p_2)(wL + rK), \; j = 1, 2 \tag{2.7}$$

---

[2] The term household is used instead of the consumer to reinforce the point that resource endowments are not owned by firms.

Given homothetic preferences, the community indirect utility function and Marshallian demands are simple linear aggregates of individual consumer welfare and demand. The marginal utility of income $v(p_1, p_2)$, and the good specific income effect $q^j(p_1, p_2)$ are common to all households.

2.2.2 *The price taking firm*

As in Section 2.1.2, both sectors are composed of a large number of identical, atomistic firms. All firms face the same input and output prices. Let y_j be the output of a firm in sector j and let $\boldsymbol{v}^j = (\ell_j, k_j) \in \mathbb{R}^2_{++}$ represent the level of labor ℓ_j and capital k_j employed by the firm. The technology for sector $j = 1, 2$ is represented by the production function $f^j : \mathbb{R}^2_{++} \to \mathbb{R}_+$, defined as $y_j = f^j(\ell_j, k_j)$, where $f^j(\cdot)$ satisfies Assumption 2. Recall from the previous discussion that production of either output requires a strictly positive level of capital and labor.

Inputs are chosen to maximize profits. Each firm can be viewed as maximizing profits in two steps. First, it chooses the input bundle (ℓ_j, k_j) that minimizes the cost of producing y_j units of output. The corresponding cost function is given by

$$C^j(w, r)\, y_j \equiv \min_{\ell_j, k_j} \left\{ w\ell_j + rk_j : y_j \leq f^j(\ell_j, k_j) \right\}, \quad j = 1, 2$$

and satisfies conditions **C1** – **C6**. In the second step, given the cost function $C^j(\cdot)\, y_j$, the firm solves the optimization problem

$$\Pi^j(p_j, w, r) \equiv \max_{y_j} \left\{ p_j y_j - C^j(\cdot)\, y_j \right\}$$

The optimal choice of y_j must satisfy the following complementary slackness condition

$$y_j \geq 0; \; p_j - C^j(\cdot) \leq 0; \; \text{and} \; \left[p_j - C^j(\cdot) \right] y_j = 0$$

Hence, in a competitive equilibrium only zero profits are possible.

2.2.3 Characterization of equilibrium

Restricting our analysis to the case where both sectors are open, i.e. $Y_1, Y_2 > 0$, equilibrium is defined by a set of factor prices and output levels $(w, r, Y_1, Y_2) \in \mathbb{R}^4_{++}$ satisfying the following four conditions:
Firms earn zero profits in each output market,

$$C^1(w,r) - p_1 = 0 \tag{2.8}$$
$$C^2(w,r) - p_2 = 0 \tag{2.9}$$

Labor and capital markets clear,

$$\sum_{j=1}^{2} C^j_w(w,r)\, Y_j = L \tag{2.10}$$
$$\sum_{j=1}^{2} C^j_r(w,r)\, Y_j = K \tag{2.11}$$

Expressions (2.8) and (2.9) require that the marginal cost of production in sector j be equal to the per-unit output price for the sector. Expression (2.10) ensures the aggregate demand for labor from the two sectors is equal to the endowment of labor L. Likewise, expression (2.11) ensures the capital market clears.

In principle, since (2.8) and (2.9) consists of two equations in the unknowns w and r, the solution may be written as

$$w = W(p_1, p_2) \tag{2.12}$$
$$r = R(p_1, p_2) \tag{2.13}$$

Notice that endowments do not appear as arguments in these equations. This result obtains because the number of traded goods equals the number of endowed factors of production.

Substituting (2.12) and (2.13) into the factor market clearing conditions (2.10) and (2.11) yields two linear equations with input-output coefficients $C^j_i(W(p_1,p_2), R(p_1,p_2))$, $i,j = 1,2$, and unknowns Y_1 and Y_2. The system is linear because the prices p_1, p_2 are exogenous in which case $C^j_i(\cdot)$ is a scalar value. Assuming both sectors produce at positive levels, denote the solution to the resulting system as

$$Y_j = Y^j(p_1, p_2, L, K), \quad j = 1, 2 \tag{2.14}$$

When both sectors produce at positive levels, and the elasticity of factor substitution between L_j and K_j is the same for both technologies, then the solution w^*, r^* satisfying the zero profit conditions is unique. Furthermore, it follows from (2.5) that $W(\cdot)$ and $R(\cdot)$ are homogeneous of degree one in p_1 and p_2,[3] while the supply functions (2.14) are homogeneous of degree zero in prices, and of degree one in endowments L and K.

The factor rental rate equations (2.12), (2.13) and the supply functions (2.14) can each be used to determine the gross domestic product function

$$\begin{aligned} G(p_1, p_2, L, K) &= W(p_1, p_2) L + R(p_1, p_2) K \qquad (2.15) \\ &= p_1 Y^1(p_1, p_2, L, K) + p_2 Y^2(p_1, p_2, L, K) \end{aligned}$$

Equation (2.15) indicates that in this model, GDP measured via the cost of production or via the value of output, yields the same result. As noted in the previous section, we can also derive the aggregate GDP function by maximizing aggregate revenue given technology (2.3) and the endowment constraints. In this case we have

$$G(p_1, p_2, L, K) \equiv \max_{L_1, L_2, K_1, K_2} \left\{ \sum_{j=1}^{2} p_j \mathcal{F}^j(L_j, K_j) : L \geq \sum_{j=1}^{2} L_j, \ K \geq \sum_{j=1}^{2} K_j \right\} \tag{2.16}$$

where $G(\cdot)$ satisfies conditions G1 – G4. If expression (2.16) is continuously differentiable, the following envelope properties apply

$$\begin{aligned} W(p_1, p_2) &= G_L(p_1, p_2, L, K) \\ R(p_1, p_2) &= G_K(p_1, p_2, L, K) \end{aligned}$$

---

[3] This result follows intuitively from the fact that $C^j(w, r) Y_j \equiv W(p_1, p_2) L_1 + R(p_1, p_2) K_1$ where the right hand expression is obviously homogenous of degree one in w and r.

and

$$Y^j(p_1, p_2, L, K) = G_{p_j}(p_1, p_2, L, K) \qquad (2.17)$$

Given the factor rental rate and supply functions, the excess demand for good j is expressed as

$$XD_j(p_j) \equiv$$

$$Y^j(p_1, p_2, L, K) - q^j(p_1, p_2) G(p_1, p_2, L, K) \begin{array}{ll} > 0 & \text{export} \\ < 0 & \text{import} \end{array}$$

Since households spend all factor income on goods, together with (2.15), Walras' law requires the value of exports to equal the value of the economy's imports,

$$\sum_{j=1}^{2} p_j XD_j(p_j) = 0$$

2.2.4 Comparative statics

The Stopler-Samuelson and the Rybczynski theorems summarize the key comparative static results of the HOS model. Stopler-Samuelson establishes the relationship between the change in the price of output and the change in factor rental rates, while Rybczynski establishes the relationship between a change in factor endowments and the change in output. These theorems apply to the two good and two endowment case, and tend to break down for more general cases. Nevertheless, the basic insights they provide can be extended, in part, to cases where the number of traded goods and number of factor endowments exceed two, as well as when the number of traded goods are greater or less than the number of factors.

The Stopler-Samuelson theorem

The theorem states, if there is an increase in the relative price of a good, then the factor used intensively in the production of that good will experience an increase in real income, while the other factor will suffer a loss in real income. In other words, an increase in the price of good j will lead to an increase (decrease) in the rental rate of the factor used intensively (extensively) in its production. Furthermore, the theorem also states, the increase (decrease) in the factor rental rate will be in greater proportion than the change in the relative price of sector j's output.

The theorem is proven in two steps. The first step shows that an increase in the price of good j causes the rental rate of the factor used intensively in its production to increase. The second step shows the percent increase in the rental rate is greater than the corresponding increase in output price. We next provide a definition of relative factor intensity. Let $j, j^* = 1, 2$, $j \neq j^*$.

Definition 1 *Sector j is capital intensive if the ratio of the profit maximizing level of K_j to L_j employed in producing good j is greater than the corresponding profit maximizing level of K_{j^*} to L_{j^*} employed in producing good j^**

$$\frac{K_j}{L_j} - \frac{K_{j^*}}{L_{j^*}} > 0 \Rightarrow S_{Kj} > S_{Kj^*}$$

where S_{Kj} denotes the share of factor K_j in the total cost of producing output j.[4]

Establishing the Stopler-Samuelson theorem only requires manipulating the zero profit conditions (2.8) and (2.9). Totally differentiating expressions (2.8) and (2.9), and manipulating the resulting expressions yields

$$C_w^j(\cdot)Y_j w\tilde{w} + C_r^j(\cdot)Y_j r\tilde{r} = p_j Y_j \tilde{p}_j, \; j = 1, 2 \tag{2.18}$$

where "$\tilde{\ }$" denotes proportional changes, e.g., $\tilde{w} = dw/w$ and $\tilde{p}_j = dp_j/p_j$. The zero profit conditions require total revenue $p_j Y_j$ be exactly equal to total cost TC_j. Dividing expression (2.18) by TC_j yields

$$\frac{wC_w^j(\cdot)Y_j}{TC_j}\tilde{w} + \frac{rC_r^j(\cdot)Y_j}{TC_j}\tilde{r} = \frac{p_j Y_j}{TC_j}\tilde{p}_j, \; j = 1, 2$$

[4] If technology is constant returns to scale Cobb-Douglas, then profit maximization implies $\alpha_{ij} = S_{ij} = w_i v_{ij}/TC_j$ where α_{ij} is the input elasticity of the i-th factor employed in the production of the j-th output.

or

$$S_{Lj}\tilde{w} + S_{Kj}\tilde{r} = \tilde{p}_j, \qquad j = 1, 2 \tag{2.19}$$

where $S_{Lj} = wC_w^j(\cdot)Y_j/TC_j$ is the (factor) cost share of labor in producing output j and $S_{Kj} = rC_r^j(\cdot)Y_j/TC_j$ is the cost (factor) share of capital in producing that output. Given zero profit, $p_jY_j = TC_j$.

Expression (2.19) is a set of two equations expressed in terms of factor shares and proportional factor rental rates $\tilde{w}$ and $\tilde{r}$. Solving this system for the proportional change in factor rental rates yields the following two equations

$$\tilde{w} = \frac{S_{K2}\tilde{p}_1 - S_{K1}\tilde{p}_2}{D_s} \tag{2.20}$$

$$\tilde{r} = \frac{-S_{L2}\tilde{p}_1 + S_{L1}\tilde{p}_2}{D_s}, \tag{2.21}$$

where

$$D_s \equiv S_{L1}S_{K2} - S_{L2}S_{K1} = S_{L1}S_{L2}\left(\frac{S_{K2}}{S_{L2}} - \frac{S_{K1}}{S_{L1}}\right)$$

$$= S_{L1}S_{L2}\frac{r}{w}\left(\frac{K_2}{L_2} - \frac{K_1}{L_1}\right)$$

By Equations (2.20) and (2.21), the sign of $\partial\tilde{w}/\partial\tilde{p}_j$ and $\partial\tilde{r}/\partial\tilde{p}_j$ each depend on the sign of D_s. If sector 2 is capital intensive, then by definition $S_{K2} > S_{K1}$, implying $K_2/L_2 > K_1/L_1$ and D_s is positive. It follows that $\partial\tilde{w}/\partial\tilde{p}_1 > 0$, while $\partial\tilde{r}/\partial\tilde{p}_1 < 0$, and conversely for a change in $\tilde{p}_2$. On the other hand, if sector 2 is labor intensive, then $S_{K2} < S_{K1}$ and D_s is negative. In this case, $\partial\tilde{w}/\partial\tilde{p}_1 < 0$, while $\partial\tilde{r}/\partial\tilde{p}_1 > 0$, and conversely for a change in $\tilde{p}_2$. Hence, an increase in the price of output j causes the rental rate of the factor used intensively in its production to rise, while the rental rate of the factor used intensively in producing output j^* falls.

Next, observe that the factor rental equations (2.12) and (2.13) are homogeneous of degree one in prices. Using Euler's theorem, and expressing the expressions in elasticity terms yields

$$\varepsilon_{p_1}^i + \varepsilon_{p_2}^i = 1, \; i = w, r \tag{2.22}$$

where $\varepsilon^i_{p_j}$ is the price elasticity of input i with respect to the price of output j. For example, the elasticity of w with respect to p_1 is equal to $\varepsilon^w_{p_1} \equiv W_{p_1}(\cdot)(p_1/w)$. Again, if sector 2 is capital intensive, then a change in the price of good 1 leads to an increase in the wage rate and a decrease in the rate of return to capital. It follows from (2.22) that one of the elasticities is negative, and consequently one elasticity in each equation must be greater than one, i.e., $\partial\tilde{w}/\partial\tilde{p}_1 > 0$ and $\partial\tilde{r}/\partial\tilde{p}_1 < 0$, implying $\varepsilon^w_{p_2} < 0$ and $\varepsilon^r_{p_1} > 1$. Thus, if sector 2 is capital intensive, $\tilde{p}_1 > 0$ implies w will increase in greater proportion than the increase in p_1. The capital rental rate r declines.

Rybczynski Theorem

The Rybczynski Theorem establishes that if the endowment of a factor increases, then the industry which uses that factor relatively intensively will (a) expand, and (b) expand more than proportionately to the percentage increase in the endowment – the other industry will contract (Woodland, 1982, p. 83). The factor market clearing conditions (2.10) and (2.11) are used to show (a) and (b).

To establish the result of the Rybczynski theorem, first substitute the factor rental rate equations (2.12) and (2.13) into the factor market clearing equations and express the result in terms of two linear equations in the endogenous variables Y_1 and Y_2

$$\begin{aligned} B_{L1}Y_1 + B_{L2}Y_2 &= L \\ B_{K1}Y_1 + B_{K2}Y_2 &= K \end{aligned}$$

where B_{Lj} and B_{Kj} are input-output coefficients for sector j and, as noted above, defined as

$$\begin{aligned} B_{Lj} &\equiv C^j_w\left(W\left(p_1, p_2\right), R\left(p_1, p_2\right)\right), \; j = 1, 2 \\ B_{Kj} &\equiv C^j_r\left(W\left(p_1, p_2\right), R\left(p_1, p_2\right)\right), \; j = 1, 2 \end{aligned}$$

Solving this system yields the supply functions

$$Y_1 = \frac{B_{K2}L - B_{L2}K}{D_B} \tag{2.23}$$

$$Y_2 = \frac{-B_{K1}L + B_{L1}K}{D_B} \tag{2.24}$$

where

$$D_B \equiv B_{L1}B_{L2}\frac{r}{w}(K_2/L_2 - K_1/L_1)$$

The sign of $\partial Y_j/\partial L$ and $\partial Y_j/\partial K$ depends on the sign of D_B, which in turn depends on the relative factor intensity term, $K_2/L_2 - K_1/L_1$. If sector 2 is relatively capital intensive, then D_B is positive and $\partial Y_1/\partial L > 0$ and $\partial Y_2/\partial L < 0$, while $\partial Y_1/\partial K < 0$ and $\partial Y_2/\partial K > 0$. This establishes part (a) i.e., one sector will expand and the other will contract.

To establish that the industry which uses that factor relatively intensively will expand more proportionately than the proportionate increase in the endowment, appeal to the linear homogeneity properties of the supply functions (2.14). Given the supply functions are homogeneous of degree one in L and K, by Euler's theorem, endowment elasticities sum to unity

$$\varepsilon_L^{y_j} + \varepsilon_K^{y_j} = 1, \ j = 1, 2 \tag{2.25}$$

where $\varepsilon_i^{y_j}$ is the sector j output elasticity with respect to endowment $i = L, K$. For example, the elasticity of sector j output with respect to labor is equal to $\varepsilon_L^{y_j} \equiv Y_L^j\left(p_1, p_2, L, K\right)\left(L/Y_j\right)$. As with Stopler-Samuelson, for each equation in (2.25), one term must be negative, and the other positive and greater than one. Hence, the sector employing the factor intensively will expand more than proportionately to the increase in this factor's endowment, while the other industry will contract. This simple argument is left to the reader as an exercise.

2.3 Generalizing the basic model

Altering the dimensions of the basic model affect whether the zero profit conditions are sufficient to solve for factor prices as functions of traded good prices alone, and moreover, whether the solution is unique. Since the dynamic models discussed in later chapters are of various dimensions, this section generalizes aspects of the equilibrium conditions of the HOS model just presented. Let M_t denote the number of traded goods produced.

2.3.1 *The case where* $M_t = N$

Suppose first, that $M_t = N$ goods are produced. The zero profit conditions are

$$C^j(\mathbf{w}) = p_j, \; j = 1, \cdots, M_t \tag{2.26}$$

and factor market clearing requires

$$\sum_{j=1}^{M_t} C^j_{w_i}(\mathbf{w}) Y_j = V_i, \; i = 1, \cdots, N \tag{2.27}$$

In principle, the system (2.26) can be used to determine the factor rental rates w_i for each factor, $i \in I$. As with the 2×2 case, the equilibrium rental rates are independent of the factor endowments. Similar to the 2×2 case, this solution can be used to determine the equilibrium output levels for each sector by substituting the solution to (2.26) into (2.27) to determine the sectoral supplies as functions of prices and endowments.[5]

2.3.2 *The case where* $M_t < N$

Assume all M_t goods are produced and all N factors are employed. Then the entire system (2.26) and (2.27) of M_t+N equations is required to solve for the endogenous variables (w_i, Y_j) $i = 1, \cdots, N$ and $j = 1, \cdots, M_t$. Denote the result by the following rental rate functions

$$w_i = W^i(\mathbf{p}, \mathbf{V})$$

and supply functions

$$Y_j = Y^j(\mathbf{p}, \mathbf{V})$$

[5] A change in the endowment of factors will not affect factor prices provided the number of open sectors M_t remain unchanged. That is, provided the economy remains within its so called cone of diversification.

As with Equation (2.15), express the GDP function as

$$G(\mathbf{p}, \mathbf{V}) = \sum_{j=1}^{M_t} p_j Y^j (\mathbf{p}, \mathbf{V}) \tag{2.28}$$

where (2.28) satisfies properties G1 – G5. Assuming (2.28) is differentiable,

$$\begin{aligned} w_i &= W^i(\mathbf{p}, \mathbf{V}) = G_{v_i}(\mathbf{p}, \mathbf{V}), \; i = 1, \cdots, N && (2.29) \\ Y_j &= Y^j(\mathbf{p}, \mathbf{V}) = G_{p_j}(\mathbf{p}, \mathbf{V}), \; j = 1, \cdots, M_t && (2.30) \end{aligned}$$

2.3.3 Comparative statics

These results apply for the case where $M_t \leq N$. Since (2.28) is homogeneous of degree one in prices and homogeneous of degree one in factor endowments, it follows that the factor rental rate price elasticities, defined as, $\varepsilon_{p_j}^{w_i} = W_{p_j}^i(\mathbf{p}, \mathbf{V})(p_j/w_i)$, sum to unity

$$\sum_{j=1}^{M_t} \varepsilon_{p_j}^{w_i} = 1, \; i = 1, \cdots, N \tag{2.31}$$

This result is analogous to (2.22). Likewise, the output endowment elasticities

$$\sum_{j=1}^{M_t} \varepsilon_{V_i}^{y_j} = 1, \; i = 1, \cdots, N \tag{2.32}$$

which is analogous to (2.25), also sum to unity. The supply elasticity of output j with regard to factor endowment i is defined as $\varepsilon_{V_i}^{y_j} = Y_{V_i}^j(\mathbf{p}, \mathbf{V})(V_i/Y_j)$.

While the factor rental rate elasticities in (2.31) sum to unity, it is not necessary for any term to be greater than one or for any term to be negative as was the case in HOS model. Thus, when the price of a good increases, no factor price need increase by a greater percentage than the percentage change in the output price, nor does a factor price need fall.

However, if there exists a product whose price p_j increase causes the ith factor rental rate to increase in greater proportion than the change in output price, i.e., if an elasticity $\varepsilon_{p_j}^{w_i}$ is greater than unity, then there must exist at least one other output $j' \neq j$ such that an increase in its price will cause w_i to fall, $\varepsilon_{p_{j'}}^{w_i} < 0$.

A similar result applies to the Rybczynski theorem. It is not necessary for any term in (2.32) to be greater than one, or for any term to be negative. If an endowment increases, the output of all goods could increase less than proportionately to the increase in the endowment. If one output does increase in greater proportion than the increase in endowment, then some other output must fall.

Finally, one can show the equilibrium rate of return to factor i is non-increasing in its own endowment, i.e.,

$$\frac{\partial W^i(\cdot)}{\partial V_i} = \frac{\partial^2 G(\cdot)}{\partial V_i^2} \leq 0,$$

while the equilibrium supply of output j is non-decreasing in its own price, i.e.,

$$\frac{\partial Y^j(\cdot)}{\partial p_j} = \frac{\partial^2 G(\cdot)}{\partial p_j^2} \geq 0$$

Since the rental rate functions are homogeneous of degree zero in endowments and the supply functions of degree zero in prices, their respective endowment and price elasticities sum to zero.

2.4 The special case of a home (non-traded) good

A dynamic three-sector model in which two goods are traded and one is only traded domestically is developed in later chapters. The presence of a fixed factor and a home-good are useful for studying the effects of a sector specific resource, such as land, on a country's transition to long-run equilibrium. Such models are also useful in studying the effects of (i) government deficit spending, (ii) foreign aid, and (iii) remittances from workers living abroad on relative prices and on the corresponding allocation of resources from traded to home-good production. This section considers the static version of such a model, the basic form and comparative static properties of which are used in later chapters.

2.4.1 *The environment*

The economy is small, open and competitive. It produces three final goods, agriculture, manufacturing, and the home-good, indexed respectively by $j = a, m, s$. The agricultural and manufacturing goods are traded internationally at given world prices p_a and p_m, while the price of the home good p_s is determined domestically. The economy is endowed with labor L, capital K, and land H. Capital and labor are economy-wide factors, while land is employed only in agriculture. Here, $\mathbf{V} = (K, L, H) \in \mathbb{R}^3_{++}$. As such, land is a resource specific to agriculture in the sense that its services can be rented in and out among firms in agriculture, but land is not used by firms in the other two sectors. In this case, $M = N = 3$, $M_t = 2$, and hence $M_t < N$. As before, households exchange the services of labor, capital, and land for wages w, capital rents r, and land rents π, where w, r, and π are each per-unit returns. All resulting income is used by households to purchase agricultural, manufacturing, and the home-good, denoted Q_a, Q_m, and Q_s respectively.

2.4.2 *Behavior of households and firms*

As with Section 2.2, households hold identical, homothetic preferences satisfying Assumption 1. Hence, the "community" indirect utility function is given by

$$\mathcal{V} = v\left(p_a, p_m, p_s\right)\left(wL + rK + \pi H\right),$$

and the corresponding Marshallian demand functions are:

$$Q_j = q^j\left(p_a, p_m, p_s\right)\left(wL + rK + \pi H\right), \ j = a, m, s$$

Firms within each sector are atomistic, identical, and hold technologies satisfying Assumption 2. Firms producing the manufactured and home-goods employ technology $f^j : \mathbb{R}^2_{++} \to \mathbb{R}_+$ defined as $y_j = f^j\left(\ell_j, k_j\right), \ j = m, s$. The corresponding sector level total cost functions are given by

$$TC_j = C^j\left(w, r\right) Y_j, \ j = m, s$$

Rather than specifying the corresponding cost function for firms producing the agricultural good, we use the sectoral value-added function (2.6). This approach can be shown to reduce the dimensionality of the problem and simplifies the comparative statics of the model. Represent the agricultural technology by the production function $f^a : \mathbb{R}^3_{++} \to \mathbb{R}_+$, defined as $y_a = f^a(\ell_a, k_a, h)$, where ℓ_a, k_a, and h are the respective levels of labor, capital, and land employed by an agricultural firm. Given the land endowment H is fixed for the sector, and given f^a is linearly homogeneous in all inputs, the sectoral aggregate technology, denoted $\mathcal{F}^a(\cdot)$, exhibits decreasing returns to scale in L_a and K_a.

Define the agricultural value-added function as

$$\pi^a(p_a, w, r) H \equiv \max_{L_a, K_a} \{p_a \mathcal{F}^a(L_a, K_a, H) - wL_a - rK_a\}$$

where H is specific to the sector, and hence, not treated as a choice variable at the sector level. By Hotelling's lemma, agriculture's partial equilibrium supply function is given by

$$y^a(p_a, w, r) H = \pi^a_{p_a}(p_a, w, r) H \tag{2.33}$$

As noted in the first section, a perfectly competitive land market among producers implies that in equilibrium, the shadow price of an additional unit of land, $\pi^a(p_a, w, r)$, is equal to the land rental rate that clears the market for land among individual producers. Thus, firms in this sector earn zero profits since, in equilibrium, the value of output is exhausted by payments to factors

$$p_a Y_a = wL_a + rK_a + \pi^a(p_a, w, r) H$$

2.4.3 The characterization of equilibrium

Restricting analysis to the case where all sectors are open, i.e. each $Y_j > 0$, equilibrium is defined by the positive values

$$(w, r, p_s, Y_m, Y_s) \in \mathbb{R}^5_{++}$$

satisfying the following conditions: two zero profit conditions in output markets,

$$C^j(w,r) - p_j = 0, \ j = m, s$$

labor and capital market clearing

$$\begin{aligned}\sum_{j=m,s} C^j_w(w,r)\,Y_j - \pi^a_w(p_a,w,r)\,H &= L\\ \sum_{j=m,s} C^j_r(w,r)\,Y_j - \pi^a_r(p_a,w,r)\,H &= K\end{aligned}$$

and clearing of the domestic market for the home-good

$$q^s(p_a,p_m,p_s)(wL + rK + \pi H) = Y_s \tag{2.34}$$

where $\pi = \pi^a(p_a,w,r)$.

The model's endogenous variables can be obtained as follows. Similar to the HOS model, the two zero profit equations can be used to express the rate of return to capital and labor as a function of the traded good price p_m, and the home-good price p_s. Express the result as

$$w = W(p_m,p_s) \tag{2.35}$$

$$r = R(p_m,p_s) \tag{2.36}$$

We determine the value of p_s shortly. Substitute (2.35) and (2.36) into the factor market clearing equations to obtain Y_m and Y_s represented by

$$Y_j = Y^j(p_a,p_m,p_s,L,K,H), \ j = m, s \tag{2.37}$$

The supply function for agriculture can be expressed in output price alone by substituting the rental rate equations (2.35) and (2.36) into the partial equilibrium supply function (2.33)

$$Y_a = Y^a(p_a,p_m,p_s)\,H = y^a(p_a,W(p_m,p_s),R(p_m,p_s))\,H \tag{2.38}$$

GDP can be expressed as a function of factor payments using (2.35), (2.36), and (2.37) as follows,

$$G(p_a, p_m, p_s, L, K, H) = W(p_m, p_s) L + R(p_m, p_s) K + \pi^a(p_a, W(p_m, p_s), R(p_m, p_s)) H \tag{2.39}$$

or equivalently by

$$G(p_a, p_m, p_s, L, K, H) = \sum_{j=m,s} p_j Y^j(p_m, p_s, L, K) + p_a Y^a (p_a, p_m, p_s) H \tag{2.40}$$

The GDP function can also be derived from the maximization problem

$$G(p_a, p_m, p_s, L, K, H) \equiv \max_{L_j, K_j} \left\{ \sum_{j=m,s} p_j \mathcal{F}^j(L_j, K_j) + p_a \mathcal{F}^a(L_a, K_a; H) \right\} \tag{2.41}$$

subject to the resource constraints

$$L \geq \sum_{j=a,m,s} L_j, \; K \geq \sum_{j=a,m,s} K_j$$

where $G(\cdot)$ satisfies properties G1 – G5.

The remaining endogenous variable is p_s. In the home-good market clearing equation (2.34), substitute (2.39) for factor payments, and (2.37) for home-good supply Y_s and solve for p_s. We focus on the role of the home-good market in the next section.

2.4.4 Selected comparative statics

The major departure from the HOS model is the presence of the home-good market. Changes in world prices and changes in endowments have direct effects on factor rental rates and output supply that are similar to those of the HOS model. However, since these variables affect the market for home-goods, they also have indirect effects on supply and factor rental rates that are transmitted through changes in the home-good price.

The price of the home-good

As noted in discussing consumer and firm behavior, both the home-good demand function, expressed as

$$Q^s(p_a, p_m, p_s, L, K, H) \equiv q^s(p_m, p_a, p_s) G(p_a, p_m, p_s, L, K, H) \tag{2.42}$$

and the supply function,

$$Y_s = Y^s(p_a, p_m, p_s, L, K, H) \equiv G_{p_s}(p_a, p_m, p_s, L, K, H) \tag{2.43}$$

are homogeneous of degree zero in prices (p_a, p_m, p_s), and homogeneous of degree one in endowments (L, K, H).

Equate home-good demand to home-good supply, and express the resulting equation in elasticity form

$$\tilde{p}_s = \sum_{j=a,m} \varepsilon_j^P \tilde{p}_j + \sum_{i=L,K,H} \varepsilon_i^P \tilde{v}_i \tag{2.44}$$

Here, $\tilde{v}_L = dL/L$, $\tilde{v}_K = dK/K$, $\tilde{v}_H = dH/H$, and

$$\begin{aligned} \varepsilon_j^P &= P_{p_j}(\cdot)\frac{p_j}{P^s(\cdot)}, \; j = a, m \\ \varepsilon_i^P &= P_{v_i}(\cdot)\frac{v_i}{P^s(\cdot)}, \; i = L, K, H \end{aligned}$$

respectively, define the elasticities of the two traded good prices and the elasticities of the three factor endowments.

In the Appendix we show the home-good price is: (i) homogeneous of degree one in traded good prices, implying

$$\sum_{j=m,a} \varepsilon_j^P = 1 \tag{2.45}$$

and (ii) homogeous of degree zero in endowments, implying

$$\sum_{i=L,K,H} \varepsilon_i^P = 0 \tag{2.46}$$

By (2.45), either ε_a^P and ε_m^P are both positive and sum to one, or one of the elasticities is negative and the other greater than one. What is the implication of one of these elasticities being negative? Let $j = m$ be the imported good. Then all else constant, an increase in p_m is referred to as a negative change in the country's terms of trade. An increase in this price can decrease real income, and "pull" more resources into production of the import competing good Y_m. In this case, both home-good supply, and demand fall, i.e., $\partial Y^s(\cdot)/\partial p_m < 0$ and $\partial Q_s/\partial p_m < 0$. If demand falls more than supply, then excess demand for the home-good declines, implying

$$\partial \tilde{p}_s/\partial \tilde{p}_m < 0 \tag{2.47}$$

in which case ε_m^P is negative. Conversely, an improvement in the country's terms of trade can cause the home-good price to increase in greater proportion than the price of the export good.

Condition (2.46) is useful in understanding the change in home-good prices in the process of economic growth. For instance, later we show if an economy's initial capital stock is less than its long-run equilibrium level, the stock of capital grows at a rate that exceeds the rate of growth in the labor force. If the home-good sector is labor intensive relative to the other two sectors, then the elasticity ε_K^P is positive, while ε_L^P is negative. In this case, as capital accumulates, the home-good price grows over time. This growth in the price of home-good dominates the negative effect of growth in the labor force. Effectively, p_s must increase in order to compete for the labor resources that are otherwise made more productive in sectors that are relatively more capital intensive than the home-good sector.

Finally, since p_s influences equilibrium factor rental rates and the equilibrium supply of manufacturing and agricultural output, it follows that changes in p_s can have *indirect* effects on these variables in the sense that home-good price effects are transmitted to (2.35), (2.36), (2.37), and (2.38) via (2.44). We now turn to these issues.

Home-good price effects on factor rental rates and supply

Since the home-good price is homogeneous of degree one in traded good prices, the factor rental rate equations (2.35) and

(2.36) remain homogeneous of degree one in traded good prices. Thus, while the rental rate elasticities sum to unity as in (2.22), the effect of a change in the price of a traded good on w and r are now more complicated. We have

$$\begin{aligned} \tilde{w} &= \varepsilon^{w}_{pm}\tilde{p}_m + \varepsilon^{w}_{ps}\tilde{p}_s && (2.48)\\ \tilde{r} &= \varepsilon^{r}_{pm}\tilde{p}_m + \varepsilon^{r}_{ps}\tilde{p}_s && (2.49) \end{aligned}$$

where the change in home-good price, $\tilde{p}_s$, is given by (2.44).

This linkage also applies to the supply functions (2.37) and (2.38). In elasticity terms

$$\tilde{Y}_j = \varepsilon^{Y_j}_{pm}\tilde{p}_m + \varepsilon^{Y_j}_{pa}\tilde{p}_a + \varepsilon^{Y_j}_{ps}\tilde{p}_s + \varepsilon^{Y_j}_{L}\tilde{L} + \varepsilon^{Y_j}_{K}\tilde{K},\ j = m, s \qquad (2.50)$$

where the land endowment H is assumed constant. The elasticities $\left(\varepsilon^{Y_j}_{pa},\ \varepsilon^{Y_j}_{pm},\ \varepsilon^{Y_j}_{ps}\right)$ are the supply response of sector j to changes in output prices p_m, p_a, and p_s. For the case of agriculture, expressing (2.33) in elasticity terms gives

$$\tilde{Y}_a = \varepsilon^{Y_a}_{pa}\tilde{p}_a + \varepsilon^{Y_a}_{w}\tilde{w} + \varepsilon^{Y_a}_{r}\tilde{r},$$

and substituting (2.48) and (2.49) into the above expression yields

$$\tilde{Y}_a = \varepsilon^{Y_a}_{pa}p_a + \varepsilon^{Y_a}_{w}\left(\varepsilon^{w}_{pm}p_m + \varepsilon^{w}_{ps}p_s\right) + \varepsilon^{Y_a}_{r}\left(\varepsilon^{r}_{pm}p_m + \varepsilon^{r}_{ps}p_s\right) \qquad (2.51)$$

Here, the elasticities $\left(\varepsilon^{Y_a}_{w}, \varepsilon^{Y_a}_{r}\right)$ are the agricultural sector's supply elasticities with respect to factor rental rates. The supply functions (2.50) and (2.51) are homogeneous of degree zero in prices, and hence the respective elasticities for each equation sum to zero.

The indirect effects in the case of (2.50) occur through the adjustment of the home-good price as determined by (2.57). In the case of Y_a, the indirect effects are transmitted through the labor and capital markets, which in turn are influenced by adjustments in the home-good price.

Traded good price effects on rental rates and supply

Let manufacturing be capital intensive relative to agriculture and the home-good, and assume manufacturing is an import competing sector. In this case, an increase in the price of manufactured goods, $\tilde{p}_m > 0$, amounts to a negative change in the country's terms of trade. It follows from Stopler-Samuelson that $\varepsilon^r_{p_m} > 0$ and $\varepsilon^w_{p_m} < 0$. Since the factor rental equations are homogeneous of degree zero, it follows that $\varepsilon^r_{p_s} < 0$ and $\varepsilon^w_{p_s} > 0$. The direct effect of $\tilde{p}_m > 0$ on rental rates is given by the elasticities $\varepsilon^w_{p_m}$, and $\varepsilon^r_{p_m}$ while the indirect effects are given by the product terms $\varepsilon^w_{p_s}\tilde{p}_s$ and $\varepsilon^r_{p_s}\tilde{p}_s$. However, $\tilde{p}_s$ is determined by the $\varepsilon^P_m\tilde{p}_m$ term in (2.44). If $\varepsilon^P_m > 0$, then in this case

$$\tilde{w} = \varepsilon^w_{p_m}\tilde{p}_m + \varepsilon^w_{p_s}\varepsilon^P_m\tilde{p}_m < 0 \tag{2.52}$$

and

$$\tilde{r} = \varepsilon^r_{p_m}\tilde{p}_m + \varepsilon^r_{p_s}\varepsilon^P_m\tilde{p}_m > 0 \tag{2.53}$$

In general, depending upon the sign of ε^P_m, the indirect effect can augment or lessen the direct effect of a change $\tilde{p}_m$ on rental rates. For instance, in the case considered here, suppose ε^P_m is negative. Then, wages fall and capital rental rates rise by a greater amount than predicted by the Stopler-Samuelson theorem.

A change in the price of the agricultural good causes a change in rental prices according to

$$\begin{aligned} \tilde{w} &= \varepsilon^w_{p_s}\left(\varepsilon^P_a\tilde{p}_a\right) \\ \tilde{r} &= \varepsilon^r_{p_s}\left(\varepsilon^P_a\tilde{p}_a\right) \end{aligned}$$

As in the case of (2.31), since the functions for w and r remain homogeneous of degree one in the prices of traded goods, when the price of a good increases, there is no need for any factor price increase to be proportionately greater than the output price increase, and no need for a factor price to fall. However, if an output price increase causes a factor rental rate to increase by a greater proportion than the change in the traded good price, then the rents to at least one factor must fall.

The effects of a change in the manufacturing price on supply (2.50) are transmitted through the terms $\varepsilon_{pm}^{Y_j}\tilde{p}_m + \varepsilon_{p_s}^{Y_j}\varepsilon_m^P\tilde{p}_m$. For $\tilde{p}_m > 0$ and $\varepsilon_m^P > 0$, manufacturing experiences a positive direct effect, $\varepsilon_{pm}^{Y_m} > 0$, and a positive indirect effect, $\varepsilon_{p_s}^{Y_m}\varepsilon_m^P\tilde{p}_m$. The home-good experiences a negative direct effect, $\varepsilon_{pm}^{Y_s}\tilde{p}_m < 0$, and a negative indirect effect, $\varepsilon_{p_s}^{Y_s}\varepsilon_m^P\tilde{p}_m < 0$. The effects on agriculture are transmitted through factor markets. However, in the case considered here, the decreasing wage has a positive effect on agricultural output while the increasing capital rental rate has a negative effect on output. The net effect depends upon the share of labor relative to capital in total cost: if agriculture is labor intensive, output can increase.

Endowment effects on rental rates and supply

The differential of (2.48), (2.49), and (2.50) with respect to endowments, can be shown to yield the following expressions:

$$\begin{aligned}
\tilde{w} &= \varepsilon_{p_s}^{w}\left(\varepsilon_L^P\tilde{L} + \varepsilon_K^P\tilde{K} + \varepsilon_H^P\tilde{H}\right) \\
\tilde{r} &= \varepsilon_{p_s}^{r}\left(\varepsilon_L^P\tilde{L} + \varepsilon_K^P\tilde{K} + \varepsilon_H^P\tilde{H}\right) \\
\tilde{Y}_j &= \varepsilon_{p_s}^{Y_j}\left(\varepsilon_L^P\tilde{L} + \varepsilon_K^P\tilde{K} + \varepsilon_H^P\tilde{H}\right) + \varepsilon_L^{Y_j}\tilde{L} + \varepsilon_K^{Y_j}\tilde{K},\ j = m, s \\
\tilde{Y}_a &= \varepsilon_w^{Y_a}\tilde{w} + \varepsilon_r^{Y_a}\tilde{r}
\end{aligned}$$

each of which show the indirect effects of changes in endowments on factor rental rates and supply. Here, we utilize the endowment components of the home-good price equation (2.44).

Consider the case where manufacturing is the most capital intensive sector while the home-good sector is the most labor intensive. Then $\varepsilon_K^{Y_m}, \varepsilon_L^{Y_s} > 0$, and $\varepsilon_L^{Y_m}, \varepsilon_K^{Y_s} < 0$. For this case, as stated above, the elasticity ε_K^P is positive, and ε_L^P is negative. Now, for purpose of the growth models presented in future chapters, consider the additional condition that growth in the capital stock exceeds the growth in labor, $\tilde{K} > \tilde{L}$. In this environment, the net effect of labor and capital accumulation on growth in the price of the home-good is positive, $\tilde{p}_s > 0$.

Under these circumstances, w increases and r falls. This result implies an increase in the productivity of labor as the capital to labor ratio increases over time, while the productivity of capital falls. Manufacturing output is affected negatively by the indirect effect of an increase in the home-good price, as determined by $\varepsilon_{p_s}^{Y_m}\left(\varepsilon_L^P\tilde{L}+\varepsilon_K^P\tilde{K}\right)$. Output is affected by Rybczynski effects, one of which is negative, $\varepsilon_L^{Y_m}\tilde{L}$, and other positive $\varepsilon_K^{Y_m}\tilde{K}$. Since manufacturing is capital intensive, it is possible for the capital effects to dominate.

The home-good sector output is affected in almost the opposite way. The home-good price effect $\varepsilon_{p_s}^{Y_s}\left(\varepsilon_L^P\tilde{L}+\varepsilon_K^P\tilde{K}\right)$ is positive, while the net factor accumulation effect, as determined by $\varepsilon_L^{Y_j}\tilde{L}+\varepsilon_K^{Y_j}\tilde{K}$ can be negative. However, the net price effect can dominate the factor accumulation effect so that growth in disposable income leads to increased consumption of the home-good, albeit at a higher price of the home-good. In this way, the home-good is competing for resources allocated to the production of traded goods so that the price ratio of traded to home-goods falls.

The effect on agricultural output once again depends on not only the magnitude of changes in w and r, but also on the sector's relative factor intensity. Changes in agricultural output, and the employment of labor and capital need not be monotonic as the labor and capital variables evolve over time.

Although most of the comparative static results in this section are ambiguous, all of the effects discussed above can be measured when a structural model is fit to data. Knowledge of these effects is crucial to explaining the evolution of a modeled economy.

2.5 Appendix: determinants of home-good price

We proceed as in Chipman (2007) to confirm (2.45). Totally differentiate expressions (2.42) and (2.43)

$$\begin{aligned} dQ^s &= Q_{p_a}^s dp_a + Q_{p_m}^s dp_m + Q_{p_s}^s dp_s \\ dY^s &= Y_{p_a}^s dp_a + Y_{p_m}^s dp_m + Y_{p_s}^s dp_s \end{aligned}$$

and convert to elasticities

$$
\begin{aligned}
\frac{dQ^s}{Q^s} &= Q^s_{p_a}\frac{p_a}{Q^s}\frac{dp_a}{p_a} + Q^s_{p_m}\frac{p_m}{Q^s}\frac{dp_m}{p_m} + Q^s_{p_s}\frac{p_s}{Q^s}\frac{dp_s}{p_s} \\
&= \sum_{j=a,m} \varepsilon_j^{Q_s}\tilde{p}_j + \varepsilon_{p_s}^{Q_s}\tilde{p}_s = 0 \qquad (2.54)
\end{aligned}
$$

$$
\begin{aligned}
\frac{dY^s}{Y^s} &= Y^s_{p_a}\frac{Y^s}{p_a}\frac{dp_a}{p_a} + Y^s_{p_m}\frac{Y^s}{p_m}\frac{dp_m}{p_m} + Y^s_{p_s}\frac{Y^s}{p_s}\frac{dp_s}{p_s} \\
&= \sum_{j=a,m} \varepsilon_j^{Y_s}\tilde{p}_j + \varepsilon_{p_s}^{Y_s}\tilde{p}_s = 0 \qquad (2.55)
\end{aligned}
$$

Define

$$
\sum_{j=a,m} \varepsilon_j^{Q_s} = \sum_{j=a,m} Q^s_{p_j}\frac{p_j}{Q^s}; \quad \sum_{j=a,m} \varepsilon_j^{Y_s} = \sum_{j=a,m} Y^s_{p_j}\frac{Y^s}{p_j}
$$

Given $Q^s(\cdot)$ and $Y^s(\cdot)$ are both homogeneous of degree zero in prices, it follows that

$$
\varepsilon_{p_s}^{Q_s} = -\sum_{j=a,m} \varepsilon_{p_j}^{Q_s} \text{ and } \varepsilon_{p_s}^{Y_s} = -\sum_{j=a,m} \varepsilon_j^{Y_s}
$$

and, hence we can rewrite (2.54) and (2.55) to obtain

$$
\begin{aligned}
&\sum_{j=a,m} \varepsilon_j^{Q_s}\tilde{p}_j - \sum_{j=a,m} \varepsilon_j^{Q_s}\tilde{p}_s \\
&= \sum_{j=a,m} \varepsilon_j^{Y_s}\tilde{p}_j - \sum_{j=a,m} \varepsilon_j^{Y_s}\tilde{p}_s = 0 \qquad (2.56)
\end{aligned}
$$

Collecting terms in (2.56) and solving for $\tilde{p}_s$ gives

$$
\sum_{j=a,m} \left(\varepsilon_j^{Y_s} - \varepsilon_j^{Q_s}\right)\tilde{p}_s = \sum_{j=a,m}^{y} \left(\varepsilon_j^{Y_s} - \varepsilon_j^{Q_s}\right)\tilde{p}_j
$$

$$
\Rightarrow \tilde{p}_s = \frac{\sum_{j=a,m}^{y} \left(\varepsilon_j^{Y_s} - \varepsilon_j^{Q_s}\right)\tilde{p}_j}{\sum_{j=a,m} \left(\varepsilon_j^{Y_s} - \varepsilon_j^{Q_s}\right)} \qquad (2.57)
$$

For a uniform rate of increase in traded good prices, $\tilde{p}_j = \tilde{p}$, $j = a, m$, (2.57) becomes

$$\tilde{p}_s = \frac{\sum_{j=a,m}^{y} \left(\varepsilon_j^{Y_s} - \varepsilon_j^{Q_s}\right)}{\sum_{j=a,m} \left(\varepsilon_j^{Y_s} - \varepsilon_j^{Q_s}\right)} \tilde{p}$$

The price of the home-good thus increases by the same proportion as the increase in world prices, establishing the result that relative prices remain unchanged, i.e.,

$$\tilde{p}_s = \tilde{p}$$

which establishes the claim that the home-good price is homogeneous of degree one in prices, expression (2.45).

We next consider the effect of endowments on home-good price. Totally differentiating demand (2.42) and supply (2.43) with respect to endowments, and expressing the result in elasticity terms yields

$$\varepsilon_{p_s}^{Q_s} \sum_{z=L,K,H} \varepsilon_z^{P} \tilde{z} + \sum_{z=L,K,H} \varepsilon_z^{Q_s} \tilde{z} = \varepsilon_s^{Y_s} \sum_{z=L,K,H} \varepsilon_z^{P} \tilde{z} + \sum_{z=L,K,H} \varepsilon_z^{Y_s} \tilde{z} \tag{2.58}$$

where the elasticities are: the direct price elasticity of home-good demand

$$\varepsilon_{p_s}^{Q_s} = Q_{p_s}^s (\cdot) \frac{p_s}{Q^s (\cdot)}$$

the endowment elasticities of the home-good price

$$\varepsilon_z^{P} = P_z^s (\cdot) \frac{z}{P^s (\cdot)}, \quad z = L, K, H$$

the endowment elasticities of home-good demand, (2.42)

$$\varepsilon_z^{Q_s} = Q_z^s (\cdot) \frac{z}{Q^s (\cdot)}, \quad z = L, K, H$$

and the endowment elasticities of home-good supply, (2.43)

$$\varepsilon_z^{Y_s} = Y_z^s (\cdot) \frac{z}{Y^s (\cdot)}, \quad z = L, K, H$$

Given a proportionate change in each endowment, i.e., $\tilde{L} = \tilde{K} = \tilde{H} = z^o$, and rearranging the terms in expression (2.58) gives

$$\left(\varepsilon_{p_s}^{Q_s} - \varepsilon_s^{Y_s}\right) \sum_{z=L,K,H} \varepsilon_z^P z^o = \sum_{z=L,K,H} \left(\varepsilon_z^{Y_s} - \varepsilon_z^{Q_s}\right) z^o \tag{2.59}$$

Solving for $\sum\limits_{z=L,K,H} \varepsilon_z^P$ yields

$$\sum_{z=L,K,H} \varepsilon_z^P = \frac{\sum\limits_{z=L,K,H} \left(\varepsilon_z^{Y_s} - \varepsilon_z^{Q_s}\right)}{\varepsilon_{p_s}^{Q_s} - \varepsilon_s^{Y_s}} \tag{2.60}$$

Since both demand and supply are homogeneous of degree one in endowments, the numerator of (2.60) is zero, thus establishing (2.46).

3

The Two Sector Ramsey Model

This chapter presents the two-sector neoclassical growth model where the transition path of consumption and saving is determined by households optimizing over time, and where firms interact in a competitive market environment. The single sector version of this model can be traced to Ramsey (1928), and its refinements by Cass (1965) and Koopmans (1965). King and Rebelo (1993) study, numerically, the transitional dynamics of the model which they confront to several stylized factors of economic growth. Barro and Sala-i-Martin (2004) also provide a full treatment of the single sector model.[1]

We present the basic model and an extension. The initial model focuses on fundamentals without being encumbered by technical change and labor force growth. The extended model introduces these latter features. We begin by describing the economic environment of the modeled economy, and the corresponding model setup. Following the pattern of the previous chapter, firm technologies and household preferences – the model's primitives – are used to derive firm level cost functions and the household expenditure function. We use these indirect representations of optimized firm and household behavior to define and characterize equilibrium. The discussion then focuses on the intra-temporal and inter-temporal features of equilibrium, and similarities between features of intra-temporal equilibrium and the HOS model presented in Chapter 2 are pointed out. We then present the key comparative statics of the model and use them to provide insights into the transition paths of prices and output supplies. To strengthen these concepts, the chapter concludes with an algebraic example of a steady-state solution,

[1] A variant of the model presented here appears in Irz and Roe (2005). They show how agriculture with land as a sector specific factor contributes to growth in early stages of economic development.

T.L. Roe et al., *Multisector Growth Models*, DOI 10.1007/978-0-387-77358-2_3,

and a numerical example of the model's equilibrium transition. This chapter provides the reader with the key concepts used in the remaining chapters of the book.

3.1 The model environment

Consider a small, closed, and competitive economy which, at time $t = 0$, is endowed with $L(0)$ units of labor and $K(0)$ units of capital. The economy has two productive sectors, indexed by $j = 1, 2$. Sector-1 produces a capital good, some of which is directly consumed by households and the rest reinvested to increase the economy's stock of capital. Sector-2 produces a pure consumption good that in equilibrium clears the domestic market at price p. The price of good-1 is the numeraire price. The underlying technology for each sector satisfies constant returns to scale. Each sector requires both labor services L and physical capital K as inputs, and both capital and labor are mobile across the two sectors. Households provide the flow of services from these resources to firms in exchange for wages w and returns to capital r. Households accumulate assets through savings, denoted $\dot{A}$. In a closed economy with no foreign ownership of assets, all capital stock is owned by domestic households. It follows that in this economy the level of assets is equal to the stock of capital, $A = K$.

3.1.1 Household behavior

Current generation households behave as though they take into account the welfare and resources of their descendants. The extended immortal family structure is appropriate if parents are altruistic in providing transfers to their children who in turn provide transfers to their children.[2] Thus, the representative household of the current generation is presumed to maximize

[2] Barro (1974) shows that this specification is equivalent to a setting where individuals are connected via a pattern of intergenerational transfers motivated by altruism.

the present value of discounted inter-temporal utility $\mathcal{U}$ subject to a budget constraint defined over an infinite horizon.

Let the quantities $Q_1(t)$,and $Q_2(t)$ denote the household's time t level of good-1 and good-2 consumption, and define consumption per worker as $q_j(t) = Q_j(t)/L$, $j = 1, 2$. Represent household preferences by the function

$$\mathcal{U} = \int_0^\infty \frac{u(q_1(t), q_2(t))^{1-\theta} - 1}{1-\theta} e^{-\rho t} dt \tag{3.1}$$

where $u(\cdot)$ is presumed to satisfy Assumption 1 of Chapter 2. This instantaneous utility $u(\cdot)$ is often referred to as a felicity function. Hereafter, we omit the (t) designation unless it is needed for clarification. Throughout the book, we assume the number of workers is proportional to total population. Observe that two parameters, the time preference rate ρ, with $\rho > 0$, and the inter-temporal elasticity of substitution $1/\theta$, with $\theta > 0$, influence the households inter-temporal choices. Econometric estimates by Giovannini (1985) suggest that in low income countries, values of $1/\theta$ are less than one, implying $\theta > 1$.

At each point in time t, the representative household provides labor services in exchange for wages $w(t)$. The household owns assets $A(t)$ that can be rented out to firms as capital or loaned to other households. In return, households receive interest income r per unit of asset. The household allocates income to purchase Q_1 and Q_2 for consumption, and saves by accumulating additional assets $\dot{A}(t)$, where the "dot" signifies a time derivative, i.e., $\dot{A} = dA/dt$. In per worker terms, the representative household's budget constraint is written

$$\dot{a}(t) = \frac{d(A(t)/L)}{dt} = \frac{\dot{A}(t)}{L}$$

$$= [w(t)L + r(t)A(t) - Q_1(t) - pQ_2(t)]\frac{1}{L}$$

$$= w(t) + r(t)a(t) - \epsilon(t) \tag{3.2}$$

where $a = A/L$, and $\epsilon = q_1 + pq_2$ is total consumption expenditure per worker.

To rule out Ponzi schemes, the credit market implicitly assumes the following transversality condition holds:

$$\lim_{t\to\infty}\left\{a(t)\cdot\exp\left[-\int_0^t r(v)dv\right]\right\}\geq 0 \tag{3.3}$$

This condition ensures the present value of assets is asymptotically non-negative.

Since the household's utility function (3.1) is inter-temporally separable, it is convenient to define a time t aggregate consumption good $q\left(t\right)$, that represents the composite of per worker good-1 and -2 consumption. In such a case, the household can be viewed as maximizing utility in two steps. In the first step, the household chooses the trajectory of composite consumption levels $q\left(0\right), q\left(1\right), \cdots$, that maximize the discounted present value of utility (3.1). In the second step, the household chooses the consumption bundle $\left(q_1\left(t\right), q_2\left(t\right)\right)$ that minimizes the expenditure of attaining each $q\left(t\right)$. Savings are determined simultaneously with expenditures on aggregate consumption $q\left(t\right)$.

Intra-temporal behavior of the household

Consider first the expenditure minimization problem. Given q and relative price p, the intra-temporal consumption problem is to choose q_1 and q_2 to minimize the cost of composite consumption q

$$\epsilon = E\left(p,q\right)\equiv\min_{(q_1,q_2)}\left\{q_1+pq_2 : q\leq u(q_1,q_2),\ (q_1,q_2)>0\right\} \tag{3.4}$$

where $E(p,q)$ satisfies properties $\mathbf{E1}-\mathbf{E6}$ of Chapter 2.

By **E5**, the expenditure function is separable in prices and q, hence $E\left(p,q\right) = \mathcal{E}(p)q$. At each instant in time, the function $\mathcal{E}(p)$ represents the price (cost) index of aggregate consumption q. Assuming $\mathcal{E}(p)$ is differentiable in p, Shephard's lemma gives the Hicksian demand function for good-2

$$q_2 = \mathcal{E}_p(p)q \tag{3.5}$$

The Hicksian demand for the other good can be expressed as

$$q_1 = \mathcal{E}(p)q - pq_2$$

To make derivations appearing later in this chapter more clear, consider an example. Let the felicity function $u(q_1, q_2)$ be Cobb-Douglas $B(q_1)^{\lambda}(q_2)^{(1-\lambda)}$, where $B = \lambda^{-\lambda}(1-\lambda)^{-(1-\lambda)}$. Then the expenditure function is equal to

$$E(p, q) = \mathcal{E}(p)\, q = p^{1-\lambda} q, \tag{3.6}$$

and the Hicksian demand for good-2 is simply

$$\begin{aligned} q_2 &= \mathcal{E}_p(p)\, q = (1-\lambda) p^{-\lambda} q \\ &= \frac{(1-\lambda)\,\epsilon}{p} \end{aligned} \tag{3.7}$$

Inter-temporal behavior of the household

Given $q = u(q_1, q_2)$ and an initial stock of assets $a(0)$, the household's inter-temporal problem is:

$$\max_{q(t)} \int_0^{\infty} \frac{q(t)^{1-\theta} - 1}{1-\theta} e^{-\rho t} dt \tag{3.8}$$

subject to the flow budget constraint

$$\dot{a}(t) = w(t) + r(t)\, a(t) - \mathcal{E}(p(t)) q(t), \tag{3.9}$$

and the limitation on borrowing (3.3).

The present value Hamiltonian associated with expressions (3.8) and (3.9) is

$$\begin{aligned} J(t) &= \frac{q(t)^{1-\theta} - 1}{1-\theta} e^{-\rho t} + \\ &\quad \xi(t)\left[w(t) + r(t)\, a(t) - \mathcal{E}(p(t)) q(t)\right] \end{aligned}$$

where ξ is a co-state variable. Assuming away corner solutions, the first order necessary conditions for a maximum of $\mathcal{U}$ are

$$\frac{\partial J}{\partial q} = 0 \Longrightarrow q^{-\theta} e^{-\rho t} - \xi \mathcal{E}(p) = 0 \tag{3.10}$$

and

$$\dot{\xi} = -\frac{\partial J}{\partial a} \Rightarrow \frac{\dot{\xi}}{\xi} = -r, \tag{3.11}$$

where $\dot{\xi} = d\xi/dt$. The transversality condition is required to hold with equality

$$\lim_{t\to\infty} \{\xi(t) \cdot a(t)\} = 0$$

This restriction is a complementary slackness condition ensuring that in the limit, the value $\xi(t)$ of assets $a(t)$ is either zero (i.e., if in the limit, the household has any assets remaining, they yield no positive value) or the household has zero assets.

Eliminate the co-state variable ξ by rearranging (3.10) to obtain

$$\xi = \frac{q^{-\theta} e^{-\rho t}}{\mathcal{E}(p)}$$

Take the log of this expression,

$$\log \xi = -\theta \ln q - \rho t - \ln \mathcal{E}(p)$$

and differentiate with respect to time to obtain

$$\frac{\dot{\xi}}{\xi} = -\theta \frac{\dot{q}}{q} - \rho - \frac{\mathcal{E}_p(p)\,\dot{p}}{\mathcal{E}(p)} \frac{p}{p} \tag{3.12}$$

Substitute (3.11) into (3.12) and rearrange terms to obtain the Euler condition

$$\frac{\dot{q}}{q} = \frac{1}{\theta}\left[r - \rho - \frac{\mathcal{E}_p(p)\,p}{\mathcal{E}(p)} \frac{\dot{p}}{p}\right] \tag{3.13}$$

It is often convenient to express this condition in terms of the rate of change in expenditure per worker. Totally differentiate the expenditure equation (3.6) to get

$$\frac{\dot{\epsilon}}{\epsilon} = \frac{\mathcal{E}_p(p)\,p}{\mathcal{E}(p)} \frac{\dot{p}}{p} + \frac{\dot{q}}{q}$$

and rearrange terms

$$\frac{\dot{q}}{q} = \frac{\dot{\epsilon}}{\epsilon} - \frac{\mathcal{E}_p(p)\,p}{\mathcal{E}(p)} \frac{\dot{p}}{p} \tag{3.14}$$

Next, note that homothetic preferences imply the term $\mathcal{E}_p(p)\, p/\mathcal{E}(p)$ is equal to the share of total expenditure spent on good-2, i.e.,

$$\frac{\mathcal{E}_p(p)\, p}{\mathcal{E}(p)} \frac{q}{q} = \frac{q_2 p}{\epsilon} = 1 - \lambda, \tag{3.15}$$

and, hence, invariant with respect to time. Substitute Equations (3.14) and (3.15) into (3.13), and simplify to obtain

$$\frac{\dot{\epsilon}}{\epsilon} = \frac{1}{\theta}\left[r - \rho - (1-\lambda)(1-\theta)\frac{\dot{p}}{p}\right] \tag{3.16}$$

For the special case where

$$\lim_{\theta \longrightarrow 1} \frac{u(q_1, q_2)^{1-\theta} - 1}{1-\theta} = \log q \tag{3.17}$$

this condition simplifies to

$$\frac{\dot{\epsilon}}{\epsilon} = r - \rho \tag{3.18}$$

An interpretation of (3.18) is the household chooses an expenditure pattern such that the rate of change in consumption expenditure is just equal to the rate of return to savings r net of the rate of time preference ρ. For $r-\rho$ large (i.e., returns to capital are higher than the rate of time preference at any given point in time), then ϵ must necessarily be relatively small. Effectively, the larger is $r-\rho$, the greater is the incentive for households to save and forego current consumption. For $\theta > 1$, expression (3.16) indicates this incentive is dampened the smaller is the inter-temporal elasticity of substitution, $1/\theta$, for given changes in $\dot{p}/p$. As we show below, $\dot{p}/p$ can be either positive or negative depending upon the capital intensity of sector-2. When returns to capital are diminishing, capital accumulation causes r to fall. In the long run, as r approaches ρ, the change in expenditure $\dot{\epsilon}$ must approach zero, i.e., the household must be in a steady-state, choosing a flat expenditure pattern.

3.1.2 Production

Firms in sector j hire labor and capital services, and employ the technology specific to their respective sector. The sector level technologies are given by

$$Y_j = \mathcal{F}^j \left(L_j, K_j\right), \; j = 1, 2$$

where $\mathcal{F}^j$ satisfies Assumption 2 in Chapter 2, Y_j denotes the aggregate output of sector j, and (L_j, K_j) denotes the level of labor and capital inputs employed in the sector. Given constant returns to scale, rewrite the production functions in intensive form as follows:

$$y_j \equiv \frac{Y_j}{L} = \mathcal{F}^j \left(\frac{L_j}{L}, \frac{K_j}{L}\right) = f^j \left(l_j, k_j\right) \qquad (3.19)$$

where $l_j = L_j/L$ and $k_j = K_j/L$.

Let r^k denote the rate of return to a unit of capital that depreciates at the rate $\delta \geq 0$. Then, the net rate of return to a unit of capital is $r^k - \delta$. Since there is no risk of default in this environment, the representative household lends to firms and other households so as to equate the returns between these two alternatives. Consequently, it follows that $r^k - \delta = r$.

The cost function corresponding to technology (3.19) is defined as

$$C^j \left(w, r^k\right) y_j \equiv \min_{l_j, k_j} \left\{ w l_j + r^k k_j : y_j \leq f^j(l_j, k_j) \right\}, \; j = 1, 2 \qquad (3.20)$$

Unlike the household, whose optimization problem includes both an inter-temporal (consumption-savings) decision and an intra-temporal (consumption basket mix) decision, the firms' optimization problem is solely intra-temporal. That is, at each point in time firms choose the level of labor and capital to minimize the cost of producing a unit of output, and then they choose output levels to exhaust revenues; independent of decisions made in other periods. As with the HOS model of Chapter 2, in equilibrium, payments to labor, wL_j and capital $r^k K$, exhaust receipts from product market sales.

3.2 Equilibrium

3.2.1 Definition and characterization of equilibrium

We restrict our analysis to an equilibrium in which both sectors produce positive levels of output in each t. In this environment, households and firms take prices as given. Given initial price $p(0)$ and economy-wide endowments $\{K(0), L(0)\}$, a competitive equilibrium is a sequence of positive good-2 prices and capital stock levels $\{p(t), k(t)\}_{t\in[0,\infty)}$, household consumption plans $\{q_1(t), q_2(t)\}_{t\in[0,\infty)}$, factor rental prices $\{w(t), r(t)\}_{t\in[0,\infty)}$ for labor and capital, and production plans

$$\{y_1(t), y_2(t), k_1(t), k_2(t), l_1(t), l_2(t)\}_{t\in[0,\infty)}$$

such that at each instant t,

1. The representative household maximizes the present value of discounted utility,
2. Firms maximize profits given their technologies, yielding zero profits,
3. Markets clear for
   (a) commodities
   $$\begin{aligned} y_1(t) - q_1(t) - \dot{k}(t) - k(t)\delta &= 0 \\ y_2(t) - q_2(t) &= 0, \end{aligned}$$
   (b) labor
   $$l_1(t) + l_2(t) = 1$$
   and
   (c) capital
   $$k_1(t) + k_2(t) = k(t)$$

Since the labor endowment is normalized to unity, $l_j(t)$ are fractions for each t. As $K(t)$ evolves over time, these fractions can change, but they always sum to unity.

Next, we divide the equilibrium into two parts: intra-temporal and inter-temporal.

Intra-temporal equilibrium conditions

At every instant t, firms in each sector equate the unit marginal cost of production to the unit output price, yielding the zero profit conditions

$$C^1\left(w, r^k\right) = 1 \quad (3.21)$$
$$C^2\left(w, r^k\right) = p \quad (3.22)$$

The labor market clearing condition is

$$\sum_j C_w^j(w, r^k) y_j = 1 \quad (3.23)$$

where $C_w^j(w, r) y_j$ represents sector j's derived demand for labor l_j. Similarly, the capital market clearing condition is given by

$$\sum_j C_{rk}^j(w, r^k) y_j = k \quad (3.24)$$

where $k = K/L$ and $C_r^j(w, r^k) y_j$ is sector j's derived demand for capital per worker, k_j. Finally, because the economy is closed, we have the additional condition that the demand for good-2 must equal its supply,

$$\frac{\partial \mathcal{E}(p) q}{\partial p} = q_2 = y_2 \quad (3.25)$$

Since expressions (3.21), (3.22), (3.23), (3.24), and (3.25) contain no equations of motion, we refer to them as the intra-temporal equilibrium conditions.

Notice that Equations (3.21) and (3.22) imply factor rental rates can be expressed as a function of the price of good-2, p. Except for normalizing the price of sector-1 output to unity, these equations are analogous to the zero profit conditions in the static HOS model presented in Equations (2.8) and (2.9).

Denote the implicit solution to (3.21) and (3.22) as

$$w = W(p) \quad (3.26)$$
$$r + \delta = r^k = R(p) \quad (3.27)$$

Although the price of good-1 is numeraire, these equations have the same properties as (2.12) and (2.13) in the HOS model discussed in Chapter 2.

Analogous to the static model, substitute the zero profit conditions (3.26) and (3.27) into the factor market clearing conditions (3.23) and (3.24). Since these equations are linear in y_j, they are readily solved to obtain the per worker supply functions

$$\begin{aligned} y_1 &= y^1(p,k) && (3.28) \\ y_2 &= y^2(p,k) && (3.29) \end{aligned}$$

GDP per worker is equal to

$$w + (r + \delta)k,$$

where, using (3.26) and (3.27), the per-worker GDP function is[3]

$$G(p,k) = W(p) + R(p)k = y^1(p,k) + py^2(p,k) \qquad (3.30)$$

The next step is to find the steady-state of this system, after which we focus on the system's equations of motion.

The steady-state equilibrium

If a steady-state exists, that is, if the economy reaches a fixed point where all variables change by an arbitrarily small amount, the Euler condition (3.16) implies

$$r^{ss} = \rho$$

since $\dot{\epsilon}/\epsilon = \dot{p}/p = 0$. Given r^{ss}, we can calculate the price of good-2, denoted p^{ss}, and the level of steady-state wage, denoted w^{ss}. Then, using the market clearing condition for good-2, Equation (3.25), the household's budget constraint can be expressed as a function of a single remaining endogenous variable, capital per worker, k.

[3] The supply function $y^2(p,k)$ could also be obtained as $\partial G(p,k)/\partial p$ and $y^1(p,k) = G(p,k) - py^2(p,k)$.

To proceed, substitute ρ into (3.27) to find the steady-state price of good-2 by solving

$$\rho + \delta = R\left(p^{ss}\right) \Rightarrow p^{ss} = R^{-1}\left(\rho + \delta\right) \tag{3.31}$$

Knowing p^{ss}, substitute it into (3.26) to obtain the level of steady-state wage

$$w^{ss} = W(p^{ss}) \tag{3.32}$$

Using $a = k$, and substituting (3.31) and (3.32) into the budget constraint (3.9) yields

$$\dot{k} = W\left(p^{ss}\right) + k\left[R\left(p^{ss}\right) - \delta\right] - \epsilon \tag{3.33}$$

in unknowns, k and expenditure ϵ.

To eliminate ϵ from (3.33), recognize that the market condition (3.25) for good-2 can be expressed as

$$\epsilon = \frac{py^2\left(p, k\right)}{1 - \lambda} \tag{3.34}$$

where we use (3.15) and the supply function (3.29). Substituting (3.34) for ϵ into (3.33) yields the steady-state budget constraint equation as a linear function of k alone,

$$\dot{k} = W\left(p^{ss}\right) + k\left[R\left(p^{ss}\right) - \delta\right] - \frac{p^{ss}y^2\left(p^{ss}, k\right)}{1 - \lambda} \tag{3.35}$$

where $\dot{k} = 0$ if a steady-state exists. The solution to (3.35) yields the steady-state capital stock per worker, k^{ss}. Given w^{ss}, r^{ss}, p^{ss}, and k^{ss}, the steady-state values of the remaining endogenous variables can be calculated from the intra-temporal conditions. Later in this chapter, we obtain an algebraic solution to k^{ss}.

Inter-temporal equilibrium conditions

The intra-temporal conditions (3.21), (3.22), (3.23), (3.24), and (3.25) consist of five equations in six unknowns (w, r, y_1, y_2, p, k). If the sequence $\{p(t), k(t)\}_{t\in[0,\infty)}$, were known, their values could be substituted into Equations (3.26), (3.27), (3.28), and (3.29) to determine the sequence of endogenous variables

$$\{w\left(t\right), r\left(t\right), y_1\left(t\right), y_2\left(t\right)\}_{t\in[0,\infty)}$$

To derive $\{p(t), k(t)\}_{t\in[0,\infty)}$, consider a system of two differential equations in k and p:

$$\dot{k} = g^k(k,p) \tag{3.36}$$
$$\dot{p} = g^p(k,p) \tag{3.37}$$

whose solution, in principle, is

$$k(t) = \mathbf{K}(k(0), p(0), t) \tag{3.38}$$
$$p(t) = \mathbf{P}(k(0), p(0), t) \tag{3.39}$$

The initial price of good-2 is $p(0)$ and $k(0)$ is the initial level of capital stock per worker. The solution (3.38) and (3.39) is a sequence $\{k(t), p(t)\}_{t\in[0,\infty)}$.

This solution, in general, can only be obtained numerically. The time elimination method is used later to solve this system. However, we defer discussion of the method and the corresponding computer code to Chapter 9. Instead, we next focus on the derivation of Equations (3.36) and (3.37).

To derive Equation (3.36), substitute the factor rental rate equations (3.26), (3.27), and the commodity market clearing equation (3.34) into the budget constraint to obtain the differential equation for $\dot{k}$ as a function of the level variables p and k,

$$\dot{k} = W(p) + k(R(p) - \delta) - \frac{py^2(p,k)}{1-\lambda} \tag{3.40}$$

To derive (3.37), time differentiate the commodity market clearing equation for good-2, (3.34) to obtain

$$(1-\lambda)\dot{\epsilon} = \left[y^2(p,k) + py^2_p(p,k)\right]\dot{p} + py^2_k(p,k)\dot{k} \tag{3.41}$$

To eliminate $\dot{\epsilon}$ from this expression, we proceed in two steps. First, replace ϵ in the Euler condition (3.16) using (3.34), and then, rearrange the result to obtain

$$\dot{\epsilon} = \frac{1}{\theta}\frac{py^2(p,k)}{1-\lambda}\left[R(p) - \delta - \rho - (1-\lambda)(1-\theta)\frac{\dot{p}}{p}\right]$$

where we have replaced r with (3.27). Second, substitute this expression for $\dot{\epsilon}$ in (3.41), and solve the result for $\dot{p}$ to obtain

$$\dot{p} = \frac{\left[R(p) - \delta - \rho\right] py^2(p,k) - \theta p y_k^2(p,k)\,\dot{k}}{\theta\left[y^2(p,k) + p y_p^2(p,k)\right] + y^2(p,k)(1-\lambda)(1-\theta)} \tag{3.42}$$

The equation is completed by replacing $\dot{k}$ by (3.40). Notice that if a steady-state exists, then $R(p) - \delta = \rho$ and $\dot{k} = 0$, and hence, $\dot{p} = 0$. It follows that the steady-state is a member of the transition path equilibria. For $\theta \to 1$, and hence preferences of the form (3.17), Equation (3.42) reduces to

$$\dot{p} = \frac{\left[R(p) - \delta - \rho\right] py^2(p,k) - p y_k^2(p,k)\,\dot{k}}{y^2(p,k) + p y_p^2(p,k)}$$

The derivation of these equations remains surprisingly unchanged throughout the book. The supply functions and the differential equations become more complicated when consideration is given to greater economy detail, like including intermediate factors of production, composite capital, and government expenditures. Nevertheless, the basic strategy for characterizing equilibrium and deriving the reduced forms of the differential equations remain essentially unchanged from one model to the next.

3.2.2 *Selected comparative statics*

This section uses results from Chapter 2 to examine the evolution of output price and production over time.

The path of prices

We maintain the assumption that both sectors of the economy produce at strictly positive levels of output, and consider the case where the economy is in transition growth, transitioning from an initial capital stock $k(0) < k^{ss}$. We state without proof, that when $k(0) < k^{ss}$, and $k(t)$ increases monotonically toward k^{ss}, then necessarily $\dot{r}/r \leq 0$. In other words, the growth in capital stock causes a decline in the marginal product of capital.

In this case, the transition path of the price of good-2 depends on factor intensity.

Claim 1 *Given $k(0) < k^{ss}$ and $\dot{r}/r \leq 0$, the transition path for the price of good-2, $\{p(t)\}_{t\in[0,\infty)}$, depends on the relative factor intensity of sector-2 production.*

Proof. The zero profit conditions (3.26) and (3.27) include the technology parameters of the economy, and thus Stopler-Samuelson-like predictions can be made. Express the returns to capital equation (3.27) in elasticity terms

$$\frac{\dot{r}}{r} = \varepsilon^r \frac{\dot{p}}{p}$$

where ε^r is the elasticity of (3.27) with respect to p. With the economy's initial condition, $k(0) < k^{ss}$ and $\dot{k} \geq 0$, diminishing returns to k imply $\dot{r}/r \leq 0$, and thus

$$\frac{\dot{r}}{r} = \varepsilon^r \frac{\dot{p}}{p} \leq 0 \tag{3.43}$$

If sector-2 production is labor intensive relative to sector-1, which implies $\varepsilon^r \leq 0$, the Stopler-Samuelson-like condition predicts the relative price p of good-2 rises, i.e.,

$$\frac{\dot{p}}{p} \geq 0 \tag{3.44}$$

On the other hand, if sector-2 production is capital intensive, then $\varepsilon^r \geq 0$ in which case

$$\frac{\dot{p}}{p} \leq 0 \tag{3.45}$$

Thus, the price p converges from below to its steady-state value p^{ss} if sector-2 production is labor intensive, and p converges from above if sector-2 production is capital intensive. ■

Claim 2 *The wage rate converges to its steady-state value from below, independent of relative factor intensity.*

Proof. The elasticity of $W(p)$ with respect to p, denoted by ε^w, takes on the opposite sign of ε^r. If sector-2 is labor intensive relative to sector-1, ε^r is negative, ε^w is non-negative, and $\dot{p}/p \geq 0$, hence

$$\frac{\dot{w}}{w} = \varepsilon^w \frac{\dot{p}}{p} \geq 0 \tag{3.46}$$

Similarly, if sector-2 is capital intensive relative to sector-1, then ε^r is positive, ε^w is negative, and $\frac{\dot{p}}{p} \leq 0$, hence

$$\frac{\dot{w}}{w} = \varepsilon^w \frac{\dot{p}}{p} \geq 0$$

Thus, independent of relative factor intensities, the wage rate always converges to its steady-state value w^{ss} from below. ■

These results tend to be surprisingly general for alternative model specifications.

The path of output supplies

Express the supply functions (3.28) and (3.29) in terms of total workers L, and assume both functions are differentiable in all arguments. Then, expressed in elasticity terms, the rate of change in each output follows:

$$\frac{\dot{Y}_j}{Y_j} = \varepsilon_p^{Y_j} \frac{\dot{p}}{p} + \varepsilon_L^{Y_j} \frac{\dot{L}}{L} + \varepsilon_K^{Y_j} \frac{\dot{K}}{K},\ j = 1, 2 \tag{3.47}$$

As shown in the previous chapter, homogeneity of degree one in factors of production implies the factor elasticities $\varepsilon_L^{Y_j}$ and $\varepsilon_K^{Y_j}$ sum to unity. For the technologies assumed, the price elasticities are $\varepsilon_p^{Y_1} < 0$ and $\varepsilon_p^{Y_2} > 0$.

From Rybczynski, if sector-j production is capital intensive, then $\varepsilon_K^{Y_j} > 0$ and $\varepsilon_L^{Y_j} < 0$, while $\varepsilon_K^{Y_{j*}} < 0$ and $\varepsilon_L^{Y_{j*}} > 0$, $j \neq j^*$. Along the transition to long-run growth, for an interval where $k(t) < k^{ss}$, it follows that $\dot{K}/K > 0$. Hence, if sector-2 is capital intensive, ε^r is positive and $\dot{p}/p$ is non-positive by **Claim 1**. In this closed economy, since demand for sector-2 output must evolve at the same rate as supply, using the expenditure function and the Euler condition (3.18), we obtain growth in sector-2 output that exceeds the difference between returns to savings and the rate of time preference

$$\frac{\dot{Y}_2}{Y_2} = r - \rho - \frac{\dot{p}}{p} \tag{3.48}$$

These results can be interpreted as follows. Since the stock of capital grows while the stock of labor is fixed, the capital intensive sector experiences a Rybczynski-like effect and tends to grow faster than the labor intensive sector. However, since supply must grow at the same rate as demand in order for the goods markets to clear, p must adjust to allow the labor intensive sector to compete for labor and capital that would otherwise be employed in the capital intensive sector. In the steady-state, (3.31) is obtained and $\dot{p}/p = 0$.

3.3 Growth in efficiency and number of workers

In this section we extend the model to include labor augmenting technological progress and population (labor force) growth. The economic environment remains unchanged, except that we allow for Harrod-neutral technological change to augment labor at a positive rate x. The level effect is expressed as $\mathcal{A}(t) = e^{xt}$. The number of workers is presumed to grow at the positive exogenous rate n, hence the time t stock of workers is equal to $L(t) = e^{nt}L(0)$, where $L(0)$ is the initial stock of workers. For notational convenience, set $L(0) = 1$. It is convenient to work with variables in quantities per unit of effective labor, where the amount of effective labor units available at any instant in time is given by $\mathcal{A}(t)L(t) = e^{(x+n)t}$. Normalizing variables by $e^{(x+n)t}$ permits the finding of a fixed point to which the endogenous variables, in units of effective workers, converge in the limit.

3.3.1 The behavior of households

The household's problem is still analogous to (3.8), however, it must now account for the net change in the number of working household members. To account for this change we normalize the household's budget constraint (3.2) by the number of household members, $L(t) = e^{nt}$, which we presume is proportional to the number of workers. This yields

$$\dot{A}(t)\,e^{-nt} = \left[w(t)\,e^{nt} + r(t)\,A(t) - \mathcal{E}(p(t))\,Q(t)\right] e^{-nt}$$

Recognizing that

$$\dot{a} \equiv \frac{dAe^{-nt}}{dt} = \dot{A}e^{-nt} - na$$

the budget constraint in per labor terms now becomes

$$\dot{a} = w + a(r-n) - \mathcal{E}(p)\,q \tag{3.49}$$

where, in per worker terms, $a = Ae^{-nt}$, $\epsilon = \mathcal{E}(p)\,q$, and $q = Qe^{-nt}$ Redefining the quantities q_1 and q_2 similarly, the household's problem is to solve

$$\max_{q_1,q_2} \int_0^\infty \frac{u(q_1(t), q_2(t))^{1-\theta} - 1}{1-\theta} e^{(n-\rho)t} dt \tag{3.50}$$

subject to (3.49), given the initial level of savings $a(0)$, and the limitation on borrowing

$$\lim_{t\to\infty}\left\{a(t)\cdot\exp\left[-\int_0^t [r(v)-n]\,dv\right]\right\} \geq 0$$

The corresponding present value Hamiltonian is

$$\begin{aligned} J &= \frac{q(t)^{1-\theta} - 1}{1-\theta} e^{(n-\rho)t} \\ &\quad + \xi(t)\left[w(t) + (r(t)-n)a(t) - \mathcal{E}(p(t))\,q(t)\right] \end{aligned}$$

Assuming away corner solutions, the first-order necessary conditions for maximizing the present value of discounted intertemporal utility are

$$\frac{\partial J}{\partial q} = q^{-\theta}e^{(n-\rho)t} - \xi\mathcal{E}(p) = 0 \tag{3.51}$$

$$\dot{\xi} = -\frac{\partial J}{\partial a} \Rightarrow \frac{\dot{\xi}}{\xi} = -(r-n) \tag{3.52}$$

and the transversality condition is

$$\lim_{t \longrightarrow \infty} [\xi(t) \cdot a(t)] = 0$$

As in the previous problem, rearrange (3.51) to obtain

$$\xi = \frac{q^{-\theta} e^{(n-\rho)t}}{\mathcal{E}(p)} \tag{3.53}$$

Then, log-differentiate (3.53) and use (3.52) to eliminate $\dot{\xi}/\xi$ to obtain the Euler condition

$$\frac{\dot{q}}{q} = \frac{1}{\theta}\left[r - \rho - (1-\lambda)\frac{\dot{p}}{p}\right] \tag{3.54}$$

where we make use of the property of homothetic preferences given in expression (3.15). This condition is identical to that of the previous problem (3.13). For the case where θ approaches 1 in the limit, the equation becomes

$$\frac{\dot{\epsilon}}{\epsilon} = \frac{\mathcal{E}_p(p)\, p}{\mathcal{E}(p)\, q}\frac{\dot{p}}{p} + \frac{\dot{q}}{q} = r - \rho \tag{3.55}$$

Since the production relationships in the following section are expressed in units per effective worker, it is useful to also express the budget constraint and the Euler condition in these terms. Normalizing the budget constraint (3.2) by $e^{(x+n)t}$ implies

$$\dot{A}e^{-(x+n)t} = \left[we^{nt} + rA - \mathcal{E}(p)\, Q\right] e^{-(x+n)t}$$

Using the notation $\hat{a} \equiv Ae^{-(x+n)t}$, $\hat{w} \equiv we^{-xt}$, and $\hat{q} \equiv Qe^{-(x+n)t}$, recognize that

$$\dot{\hat{a}} = \frac{dAe^{-(x+n)t}}{dt} = \dot{A}e^{-(x+n)t} - \hat{a}(-x-n)$$

which leads to the normalized budget constraint

$$\dot{\hat{a}} = \hat{w} + \hat{a}(r - x - n) - \mathcal{E}(p)\, \hat{q}\ , \tag{3.56}$$

expressed in units per effective worker.

To express the Euler condition in expenditure per effective worker terms, log-differentiate the expenditure equation, $E(p)\,\hat{q}$,

$$\frac{\dot{\hat{\epsilon}}}{\hat{\epsilon}} = \frac{\mathcal{E}_p(p)\,p}{\mathcal{E}(p)}\frac{\dot{p}}{p} + \frac{\dot{q}}{q} - x$$

where $\dot{\hat{q}}/\hat{q} = \dot{q}/q - x$, and substitute this result for $\dot{q}/q$ in (3.54) to obtain

$$\frac{\dot{\hat{\epsilon}}}{\hat{\epsilon}} = \frac{1}{\theta}\left[r - \rho - \theta x - (1-\lambda)(1-\theta)\frac{\dot{p}}{p}\right] \tag{3.57}$$

For the case of unitary elasticity of inter-temporal substitution, (3.17), we obtain

$$\frac{\dot{\hat{\epsilon}}}{\hat{\epsilon}} = r - \rho - x \tag{3.58}$$

In the long-run, if a steady-state exists, it must be the case that $\dot{\hat{\epsilon}}/\hat{\epsilon} = 0$. This result implies the rate of expenditure per worker in the long-run is positive forever, and

$$\frac{\dot{\epsilon}}{\epsilon} = x$$

Technological progress precludes diminishing returns to capital in the long-run. In contrast to (3.16) and (3.18), the presence of exogenous growth in effective labor supply at the rate x causes the household's steady-state rate of return r^{ss} to exceed the rate of time preference ρ.

3.3.2 Production

As in the case with $x = 0$, firms still employ technologies $f^j(\cdot)$ satisfying Assumption 2 of Chapter 2. Now, however, we normalize output, capital, and labor as follows: Let $\hat{y}_j(t) = Y_j(t)e^{-(x+n)t}$, $\hat{k}_j(t) = K_j(t)\,e^{-(x+n)t}$ and $l_j(t) = L_j(t)/L(t)$. Then, the technology for sector j in intensive form is

$$\hat{y}_j = \mathcal{F}^j(l_j, \hat{k}_j),$$

and satisfies Assumption 2.2. The behavior of a firm is analogous to the previous section, with cost functions given by

$$C^j(\hat{w}, r^k)\hat{y}_j \equiv \min_{l_j,\hat{k}_j} \left\{ \hat{w} l_j + r^k \hat{k}_j : \hat{y}_j \leq \mathcal{F}^j(l_j, \hat{k}_j) \right\} \; j = 1, 2 \tag{3.59}$$

where $C^j(\cdot)\,\hat{y}_j$ satisfies properties **C1** – **C6** of Chapter 2. Shephard's lemma gives the sector j labor demand function

$$C^j_{\hat{w}}(\hat{w}, r^k)\hat{y}_j = l_j, \quad j = 1, 2$$

and the sector j capital demand per economy-wide effective worker[4]

$$C^j_{r^k}(\hat{w}, r^k)\hat{y}_j = \hat{k}_j, \; j = 1, 2$$

3.3.3 Equilibrium

The definition and characterization of equilibrium remain essentially unchanged from that of the previous section, except the endogenous variables are expressed in per effective worker terms. Since the cost functions are expressed in terms of the wage rate per effective worker, the reduced form factor rental rate equation (3.26) is expressed in terms of the normalized wage rate, $\hat{w} = W(p)$.

Following the same procedure used to derive expression (3.40), substitute the reduced forms for $\hat{w}$, and r, and the market clearing condition

$$\hat{\epsilon} = \mathcal{E}(p)\,\hat{q} = p\frac{y^2\left(p, \hat{k}\right)}{1-\lambda}$$

into the budget constraint (3.56) to obtain the following differential equation for $\hat{k}$

$$\dot{\hat{k}} = W(p) + \hat{k}\left[R(p) - \delta - n - x\right] - p\frac{y^2\left(p, \hat{k}\right)}{1-\lambda} \tag{3.60}$$

For $r^{ss} = \rho + \theta x$, the remaining steady-state values of the endogenous variables can be calculated using the reduced form

---

[4] The envelope theorem can be used to verify these results.

factor rental rate equations, the budget constraint for $\dot{\hat{k}} = 0$, and the intra-temporal equilibrium conditions.

Following the same steps used to derive the differential equation (3.42) for p, differentiate the market clearing equation for good-2 with respect to time

$$(1-\lambda)\,\dot{\hat{\epsilon}} = \left[y^2\left(p,\hat{k}\right) + p y_p^2\left(p,\hat{k}\right)\right]\dot{p} + p y_{\hat{k}}^2\left(p,\hat{k}\right)\dot{\hat{k}} \qquad (3.61)$$

Then, use the Euler condition (3.57) and the market clearing condition for good-2 to express $\dot{\hat{\epsilon}}$ as

$$\dot{\hat{\epsilon}} = \frac{1}{\theta}\frac{p y^2\left(p,\hat{k}\right)}{1-\lambda}\left[R(p) - \delta - \rho - \theta x - (1-\lambda)(1-\theta)\frac{\dot{p}}{p}\right]$$

Substitute this equation into (3.61) and solve for $\dot{p}$ to get a result almost identical to the previous case with the exception of the term θx and the normalized variables:

$$\dot{p} = \frac{\left[R(p) - \delta - \rho - \theta x\right] p y^2\left(p,\hat{k}\right) - \theta p y_{\hat{k}}^2\left(p,\hat{k}\right)\dot{\hat{k}}}{\theta\left[y^2\left(p,\hat{k}\right) + p y_p^2\left(p,\hat{k}\right)\right] + y^2\left(p,\hat{k}\right)(1-\lambda)(1-\theta)} \qquad (3.62)$$

If a steady-state exists, then it can be seen that $\dot{p} = 0$.

3.3.4 Comparative statics

Claim 1 remains valid so that (3.43), (3.44) and (3.45) remain unchanged. In the case of **Claim 2**, since $\dot{\hat{w}}/\hat{w} = \dot{w}/w - x$, (3.46) becomes

$$\frac{\dot{w}}{w} = \varepsilon^w \frac{\dot{p}}{p} + x \geq 0$$

This result suggests the wage rate grows in the long-run at rate x.

Analogous to (3.30), the GDP function per effective worker is given by

$$G\left(p,\hat{k}\right) = W(p) + R(p)\,\hat{k}$$

Assuming differentiability, it can be shown that

$$G_p\left(p, \hat{k}\right) = y^2\left(p, \hat{k}\right)$$

and

$$y^1\left(p, \hat{k}\right) = G_p\left(p, \hat{k}\right) - py^2\left(p, \hat{k}\right)$$

Expressing the supply functions in non-normalized form, it can be shown that output evolves according to

$$\frac{\dot{Y}_j}{Y_j} = \varepsilon_p^j \frac{\dot{p}}{p} + (x+n)\,\varepsilon_{AL}^j + \varepsilon_K^j \frac{\dot{K}}{K} \tag{3.63}$$

where the elasticities, ε_{AL}^j, and ε_K^j sum to unity and represent the Rybczynski effect. In the steady-state, the rate of growth in the stock of capital is $x+n$ which equals the rate of growth in output. For the technologies considered here, ε_p^2 is positive while ε_p^1 is negative.

An algebraic and a numerical example are provided in the next two sections to provide further insights into the model. The algebraic example provides a solution to the steady-state for the case of unitary inter-temporal elasticity of substitution, and Cobb-Douglas preferences and technologies. The numerical example is based on data of the Turkish economy that we also use for other examples throughout the book.

3.4 An algebraic example

Consider the case where the felicity function $u(q_1, q_2)$ is

$$u(q_1, q_2) \equiv q = \Lambda q_1^{\lambda} q_2^{1-\lambda} \tag{3.64}$$

An inter temporal elasticity of substitution of unity is assumed. Production technologies are

$$\hat{y}_1 = \Psi_1 l_1^{\alpha} \hat{k}_1^{1-\alpha} \tag{3.65}$$

$$\hat{y}_2 = \Psi_2 l_2^{\beta} \hat{k}_2^{1-\beta} \tag{3.66}$$

To minimize notation, the scale parameters Λ, Ψ_1, and Ψ_2 are chosen such that, for the case of the expenditure function, $\Lambda \equiv \lambda^{-\lambda}(1-\lambda)^{\lambda-1}$, and for the cost functions, $\psi_1 \equiv \alpha^{-\alpha}(1-\alpha)^{\alpha-1}/\Psi_1$, $\psi_2 = \beta^{-\beta}(1-\beta)^{\beta-1}/\Psi_2$.

The zero profit conditions (3.26) and (3.27) imply[5]

$$\hat{w} = \psi_1^{\frac{1-\beta}{\alpha-\beta}}(\psi_2 p)^{\frac{\alpha-1}{\alpha-\beta}} \tag{3.67}$$

$$r = \psi_1^{\frac{-\beta}{\alpha-\beta}}(\psi_2 p)^{\frac{\alpha}{\alpha-\beta}} - \delta \tag{3.68}$$

Note that (3.68) confirms the comparative static result (3.44) when sector-1 is capital intensive, i.e, $\alpha - \beta < 0$.

Using (3.67) and (3.68), GDP per effective worker is

$$\hat{G}\left(p, \hat{k}\right) = \psi_1^{\frac{1-\beta}{\alpha-\beta}}(\psi_2 p)^{\frac{\alpha-1}{\alpha-\beta}} + \psi_1^{\frac{-\beta}{\alpha-\beta}}(\psi_2 p)^{\frac{\alpha}{\alpha-\beta}}\hat{k}$$

The supply of good-2 per unit of effective labor is

$$\hat{y}_2 = \frac{\alpha-1}{\alpha-\beta}\psi_1^{\frac{1-\beta}{\alpha-\beta}}(\psi_2)^{\frac{\alpha-1}{\alpha-\beta}}p^{\frac{\beta-1}{\alpha-\beta}} + \frac{\alpha}{\alpha-\beta}\psi_1^{\frac{-\beta}{\alpha-\beta}}(\psi_2)^{\frac{\alpha}{\alpha-\beta}}p^{\frac{\beta}{\alpha-\beta}}\hat{k} \tag{3.69}$$

Since

$$\hat{y}_1 = \hat{G}\left(p, \hat{k}\right) - p\hat{y}_2$$

we obtain

$$\hat{y}_1 = \frac{1-\beta}{\alpha-\beta}\psi_1^{\frac{1-\beta}{\alpha-\beta}}(\psi_2 p)^{\frac{\alpha-1}{\alpha-\beta}} + \frac{-\beta}{\alpha-\beta}\psi_1^{\frac{-\beta}{\alpha-\beta}}(\psi_2 p)^{\frac{\alpha}{\alpha-\beta}}\hat{k} \tag{3.70}$$

If sector-1 is capital intensive, and $\dot{K}/K > x + n$, sector-1 output is affectived positively and sector-2 output negatively. However, since the price p of the sector-2 commodity is endogenous, **Claim 1** suggests $\dot{p}/p \geq 0$. Sector-2 firms are thus able to offset the advantage sector-1 obtains from growth in capital stock.

---

[5] Only the price of good-2 appears because the price of good-1 is normalized to unity.

If a steady-state exists, then for $\theta \longrightarrow 1$,

$$r^{ss} = \psi_1^{\frac{-\beta}{\alpha-\beta}} \left(\psi_2 p\right)^{\frac{\alpha}{\alpha-\beta}} - \delta = \rho + x \tag{3.71}$$

which implies a steady-state price

$$p^{ss} = \psi_1^{\frac{\beta}{\alpha}} \left(\psi_2\right)^{-1} \left(\rho + x + \delta\right)^{\frac{\alpha-\beta}{\alpha}} \tag{3.72}$$

Substituting (3.72) into (3.67) determines steady-state wage rate

$$\hat{w}^{ss} = \psi_1^{\frac{1}{\alpha}} \left(\rho + x + \delta\right)^{\frac{\alpha-1}{\alpha}} \tag{3.73}$$

To solve for steady-state $\hat{k}$, substitute expressions (3.71) – (3.73) into the budget equation (3.60) to obtain

$$\hat{k}^{ss} = \frac{\psi_1^{\frac{1}{\alpha}} (\rho + x + \delta)^{\frac{\alpha-1}{\alpha}} \left(1 - \frac{\alpha-1}{\alpha-\beta}\frac{1}{1-\lambda}\right)}{\frac{\alpha(\rho+x+\delta)}{(1-\lambda)(\alpha-\beta)} - (\rho - n)} \tag{3.74}$$

For the special case $\delta = x = n = 0$, we get

$$k^{ss} = \frac{\left[(\alpha - \beta)\lambda + \beta - 1\right]}{(\alpha - \beta)\lambda + \beta} \left(\frac{\psi_1}{\rho}\right)^{\frac{1}{\alpha}} \tag{3.75}$$

It can be seen that the steady-state level of capital, and hence GDP per worker, decreases with an increase in the labor intensity of the economy,

$$\left(\frac{\partial k^{ss}}{\partial \alpha}, \frac{\partial k^{ss}}{\partial \beta}\right) < 0$$

The steady-state level of capital per effective worker falls (rises) with an increase (decrease) in the population growth rate and the rate of technological change

$$\left(\frac{\partial \hat{k}^{ss}}{\partial n}, \frac{\partial \hat{k}^{ss}}{\partial x}\right) < 0$$

GDP also falls with an increase in "impatience," ρ. The steady-state level of capital is increasing in the scale parameter, ψ_1, of the capital producing sector, and increasing in the share λ of income spent on the capital good.

To solve for steady-state expenditure $\hat{\epsilon}^{ss}$, substitute (3.75) into the budget constraint to get

$$\hat{\epsilon}^{ss} = \psi_1^{\frac{1}{\alpha}} (\rho + x + \delta)^{\frac{\alpha-1}{\alpha}} +$$

$$(\rho - x - n - \delta) \left[\frac{\psi_1^{\frac{1}{\alpha}} (\rho + x + \delta)^{\frac{\alpha-1}{\alpha}} \left(1 - \frac{\alpha-1}{\alpha-\beta} \frac{1}{1-\lambda}\right)}{\frac{\alpha(\rho+x+\delta)}{(1-\lambda)(\alpha-\beta)} - (\rho - n)} \right]$$

Finally, for the special case where $\delta = x = n = 0$;

$$\epsilon^{ss} = \frac{\psi_1^{\frac{1}{\alpha}} \rho^{\frac{\alpha-1}{\alpha}}}{(1-\lambda)\beta + \alpha\lambda}$$

and the steady-state level of per capita aggregate consumption is recovered using

$$q^{ss} = (p^{ss})^{\lambda-1} \epsilon^{ss}$$

$$\Rightarrow q^{ss} = \frac{\psi_1^{\frac{1-(1-\lambda)\beta}{\alpha}} (\psi_2)^{1-\lambda} \rho^{\frac{(1-\lambda)\beta+\alpha\lambda-1}{\alpha}}}{(1-\lambda)\beta + \alpha\lambda} \qquad (3.76)$$

These solutions permit the calculation of the remaining endogenous variables for the steady-state of the model.

3.5 A numerical example

We provide a numerical example for each of the general models presented in this book. Each of these numerical examples draw upon data from easily accessible sources. These are the World Bank's World Development Indicators (WDI) data base, and the Global Trade, Assistance, and Production, Version 6 (GTAP-6) data base (see Badi and Walmsley (2008)). The GTAP data are available through the Center for Global Trade Analysis, Purdue University. The economy is divided into two sectors. One sector is an aggregate of the industrial and agricultural sector, and the other is the rest of the economy which we refer to as the service sector. A discussion of the International Standard Industrial Classification (ISIC) defining the subsectors comprising these sectors is available in Chapter 8.

To generate the numerical solution, we follow closely the characterization of equilibrium laid out in this chapter. The time elimination method (discussed in Chapter 9) is used to numerically solve the two differential equations (3.60) and (3.62) which determine the trajectories of capital stock and service sector price. We then use these results and the intra-temporal conditions to obtain the numerical values of the model's remaining endogenous variables.

3.5.1 Parameter estimation

We fit the empirical model of this chapter, and all of the empirical models presented in Chapters 4, 5, 6, and 7, to year 2001 Turkish data. We use the GTAP and WDI data to estimate most of the numerical values of the model's parameters. Parameter estimates not based on these data are the preference parameters ρ, θ, and depreciation δ. The details of these procedures are discussed in Chapter 8. Table 3.1 presents the set of structural parameters common to each model.

Table 3.1 Parameter and initial values common to all examples

θ	ρ	δ	x	n	β	$K(0)$ in 2001 Lira
1.26	0.04	0.04	0.019	0.0146	0.529	6.2×10^{17}

Source: Author estimates and calculations using WDI and GTAP data

The remaining structural parameters of the two-sector model include labor's share in total factor cost of the industrial-agricultural sector, and household expenditure share for each of the two goods consumed. The labor cost share for industrial-agricultural production is $\alpha = 0.48$, and $\beta = 0.529$ for the production of the service good. The value of β is common to all numerical examples because the definition of this sector remains unchanged throughout. The share of total household expenditure allocated to the industrial-agricultural good is $\lambda = 0.345$.

The values of α and λ change in other examples as the empirical model becomes more disaggregated.

These parameters lead to a number of predictions. The value of x and n suggest the exogenous rate of long-run growth of the model's level variables is $x+n$. Since $\beta > \alpha$, the service good sector is relatively more labor intensive. For $\hat{k}(0) < \hat{k}^{ss}$, growth in the capital stock will have Rybczynski-like effects on the growth in the industrial-agricultural sector supply, while growth in labor services will affect positively the growth in supply of the service good. Since the growth rate of $K(t)$ in transition exceeds the rate of growth in labor services, the rise in price of the service good is required in order for the service good sector to compete for resources needed for the service good market to clear. Thus, as **Claim 1** predicts, the price of the service good will converge to its long-run value from below.

3.5.2 Empirical results

The model predicts a rate of growth of GDP per worker of 3.05 percent in year 2002. The economy reaches the half-way point to steady-state rate of growth in GDP of 1.9 percent per worker by the year 2013. By 2050, the economy is within 4 percent of this long-run rate. The main results are presented in Tables 3.2 and 3.3. From Table 3.2, it can be seen that the capital stock to GDP ratio is about 4 in 2001 and grows to 4.6 by 2031. Capital per worker increases by a factor of 1.4 over the 2001–2011 period, and by a factor of 2.4 by 2031. At the half-way point, GDP per worker increases by 40 percent and more than doubles by year 2031. A consequence of capital deepening is an increase in wage income per worker. The decline in the capital rental rate and rise in capital stock per worker are sufficient to cause the share of total income accounted for by wages and the capital asset to remain almost equal over time.

Changes in the value share of each sector's output in GDP and each sector's factor share is limited because the economy is closed. The share of the industrial-agricultural sector's value of output in GDP exceeds its corresponding household expenditure

Table 3.2 Factor income, expenditure and saving*

Year	GDP per worker	Capital per worker	Wage income per worker	Capital rent per worker	Expenditure per worker	Saving to GDP ratio
2001	6435	25896	3220	3215	3916	0.39
2006	7453	31466	3732	3721	4610	0.32
2011	8503	37148	4260	4243	5323	0.27
2016	9602	42995	4812	4790	6063	0.22
2021	10765	49074	5396	5369	6841	0.19
2026	12007	55457	6020	5987	7666	0.16
2031	13341	62220	6690	6651	8547	0.13

*In millions of 2001 Turkish Lira

Source: model results

share. This result obtains because some of this sector's output is an increment to the stock of capital in each period and the remainder is consumed as a final good. Consequently, during the early stages of transition growth, the industrial-agricultural sector accounts for about 60 percent of GDP and declines as the ratio of savings to GDP falls. The share converges to a constant in the long-run that is sufficient to account for the growth in the supply of effective labor, x.

The service sector's value share in GDP rises from about 40 percent in 2002 to slightly over 43 percent in the long-run. The share of the economy's labor and capital employed in each sector remain relatively constant over time, in spite of capital deepening and the rise in the wage rate. The share of capital in the industrial-agricultural sector is 63 percent initially, and falls slightly to 60 percent by 2031. The share of labor in the relatively labor intensive service sector is 42 percent in 2001, and rises to about 44 percent by 2031.

These changes can be further explained by drawing upon the Rybczynski-like effects of growth in the economy's stock of capital and effective labor, as shown in Table 3.3. The left panel reports the results of performing a growth-accounting exercise using Equation (3.63) for the industrial-agricultural sector and the right panel reports results for the service sector. The percent price and factor contributions to growth equal the total rate of growth reported in the first column of the table.

As capital deepening occurs, the most capital intensive sector, the industrial-agricultural sector, realizes a positive Rybczynski effect on growth in output from growth in the capital stock, and a negative effect from the growth in labor services. This sector's total output grows by over 4 percent in early transition, and declines to 3.5 percent by 2031 as the rate of capital deepening slows. As this sector employs a larger share of the economy's total capital services, its productivity of labor rises which makes profitable the employment of more labor at a higher wage rate. The service sector is the most labor intensive and experiences a positive Rybczynski effect on growth in output from the growth in labor services an a negative effect from the growth

Table 3.3 Output growth and factor contributions

	Industry-Agriculture sector				Service sector			
		Contributions to Growth				Contributions to Growth		
Year	Output growth	Service price	Capital stock	Effective labor	Output growth	Service price	Capital stock	Effective labor
2001	0.0433	−0.2073	0.5250	−0.2744	0.0487	0.3129	−0.7207	0.4565
2006	0.0404	−0.1480	0.4660	−0.2776	0.0443	0.2173	−0.6223	0.4493
2011	0.0384	−0.1070	0.4255	−0.2800	0.0413	0.1540	−0.5569	0.4441
2016	0.0371	−0.0780	0.3970	−0.2818	0.0392	0.1107	−0.5120	0.4405
2021	0.0361	−0.0573	0.3766	−0.2832	0.0377	0.0803	−0.4804	0.4378
2026	0.0355	−0.0422	0.3618	−0.2842	0.0366	0.0587	−0.4580	0.4358
2031	0.0350	−0.0312	0.3511	−0.2849	0.0358	0.0431	−0.4417	0.4344

Source: Model results

in capital stock (right panel of Table 3.3). However, since the growth in capital stock converges from above to the growth in labor services, if all else were constant, capital deepening in the industrial-agricultural sector would increase its marginal value product of labor relative to the growth in the marginal value product of labor in the service sector.

Due to the service sector's relative labor intensity, firms in this sector need to employ proportionately more labor than the other sector to meet the growth in service sector demand brought about by the growth in household income shown in Table 3.2. While capital deepening induces a decline in the unit cost of capital, the relatively small share of capital in the total cost of producing the service good does not provide sufficient incentives for firms to increase output to meet final demand. Consequently, the market price for the service good must increase in order for firms to complete for the labor and capital necessary to increase supply and cause the market for the service good to clear. The price of the service good, measured in terms of the numeraire price of the other good increases, albeit slightly, from unity in 2001 to 1.015 by 2031. This increase contributes positively to the growth in service sector output, and dampens the growth in output of the industrial-agricultural good (left panel). As the economy approaches its long-run equilibrium, the price effect approaches zero, and the net effect of growth in resources decline, causing growth in each sector's total output to converge to the same long-run rate of growth equal to $x + n$.

3.6 Conclusion

In this chapter we introduced the most basic two-sector growth model in which the path of consumption and saving are determined by optimizing households and firms that interact in a competitive market environment over time. We first presented the model in its simplest form in order to focus on the fundamentals of consumer optimization, and the definition and characterization of equilibrium. The basic steps of stating the model's primitives, and defining and characterizing equilibrium in a way which emphasizes the intra- and inter-temporal features of the model form a pattern that prevails throughout the book. The intra-temporal characterization is shown to closely resemble that of the two-sector static model presented in the previous chapter which links the dynamic model to the static trade theory literature and allowed us to draw upon many of the comparative static results found in the literature.

The basic model is extended to account for exogenous growth in labor services and and exogenous growth in labor supply that is assumed proporational to the growth in population. This extension shows how to normalize the model's variables in effective labor units and the changes this implies to equlibrium conditions, such as the Euler condition. The derivation of the model's two differential equations also establishes the pattern for their derivation used in the remaining chapters, although the level of complexity and number of equations increase. This chapter thus establishes the basic point of departure of the development of more complex models.

The empirical example drew directly upon the equations that characterized equilibrium, and provided numerical values for the case of Turkey of its transition path to long-run equilibrium. Chapter 8 lays out clearly how the data are organized, while Chapter 9 shows how the time elimination method is used to numerically solve the model. The more complex models developed in the remainder of the book draw upon the same but more disaggregated data permitting a comparison of the empirical results generated from each of the models, as well as providing insights into the advantages of each respective model.

A limitation of the two-sector model is that the economy is closed which is shown to limit changes in sector value shares in GDP over time as well as sectoral factor shares. The two sector – closed economy structure not only limits policy analysis to a fairly aggregate level of the economy, it also precludes the analysis of trade policy and other issues of an open economy. The next chapter builds upon the fundamentals of this chapter by developing a three-sector model of a small and open economy.

4

The Three-Sector Ramsey Model

This chapter develops a three-sector growth model with three factors of production. One factor is specific to a sector, and one sector's output is a home-good, meaning it is not traded in international markets. The chapter builds upon the static three-sector model developed in Chapter 2, and the two sector Ramsey model presented in the previous chapter. The dynamic three-sector model is a convenient point of departure for developing policy models with more sectoral detail, and for studying various other aspects of economic growth that have received attention at least from the time of Arthur Lewis. The seminal work of Lewis (1954), further developed by Fei and Ranis (1961) emphasize the supply of surplus labor from the farm sector to the rest of the economy as an essential part of the growth process. This theme was also emphasized in the work of Jorgenson (1967). In spite of the renewed interest in growth theory in the 1980s, Matsuyama (1992) was among the first to develop a model of endogenous growth with two distinct sectors, agriculture and manufacturing. In a series of papers, Echevarria (1995, 1997, 2000), and more recently, Gollin et al. (2004) develop neoclassical growth models in which agriculture and a home-good are used to show how the sectoral composition of an economy explains an important part of the variation in growth rates across countries.

In the absence of growth in factor productivity, the sector specific factor allows for diminishing returns to labor and capital to occur more rapidly than in the other sectors while the evolution of the price of the home-good relative to traded goods explains the process by which the non-traded sector competes with the traded good sectors for resources. The asset market also receives attention because capital and the sector specific

T.L. Roe et al., *Multisector Growth Models*, DOI 10.1007/978-0-387-77358-2_4,

factor are the two assets held by households. In the process of growth, the decline in the rental rate of capital and the change in the rental rate of the sector specific factor suggest the price of this factor also evolves over time. Although not pursued in this chapter, the model provides a point of departure for studying asset market failures that preclude agents from arbitraging the differences in asset yields.

The basic model is first presented and selected comparative static results are shown. The model is then modified to included Stone-Geary preferences. The appendix uses this modified model to extend Caselli and Ventura's (2000) representative consumer theory of distribution. The chapter concludes with a numerical example and provides the basis for the next chapter which extends the model to account for intermediate factors of production, composite capital, and government.

4.1 The model environment

The modeled economy is a small, open and perfectly competitive economy that produces and consumes three final goods: an agricultural, manufacturing, and a home-good, indexed, respectively, $j = a$, m and s. The economy is initially endowed with $L(0)$ and $K(0)$ units of labor and capital, and H units of land. The land endowment remains constant over time. The manufactured and the agricultural good are traded internationally at fixed prices p_m and p_a, respectively. The home-good is only traded in the domestic economy at the endogenous price p_s. The services of labor and capital are employed in the production of all three goods, while land is employed only in agricultural production. A land rental market among farmers is presumed to exist so that land can be rented at a rate Π. The manufactured good enters final consumption, and contributes to the economy's stock of capital with any excess supply or demand traded in international markets at the price p_m. The agricultural and the home-good are pure consumption goods. Labor services are not traded internationally and domestic residents own the entire stock of domestic assets.

As in Chapter 3, the number of workers is assumed to grow at rate n, and Harrod-neutral technological change augments labor at the rate x. Hence, $\mathcal{A}(t) L(t) = L(0) e^{(x+n)t}$ is the time t stock of effective labor. The initial stock of labor $L(0)$ is normalized to unity. New in this chapter is exogenous technological change in agriculture, $\mathcal{B}(t) = e^{\eta t}$, $\eta > 0$. For fixed H, the effective units of land at t is $\mathcal{B}(t) H$. Unless indicated otherwise, we assume the *sustainability condition*

$$\eta = x + n \tag{4.1}$$

The implications of this assumption are discussed later.

At each instant in time, households provide labor services in exchange for a wage $w(t)$ and earn income at rate $r(t)$ on capital and land assets $A(t)$. They consume the three goods, Q_j, $j = a, m, s$ and incur expenditures $\sum_{j=a,m,s} p_j Q_j$. Disallowing foreign ownership of assets, the total value of asset holdings can be expressed as

$$A(t) = K(t) + P_H(t) H$$

where the price p_m of capital is normalized to unity, and P_H is the price of a unit of land.[1]

4.1.1 No-arbitrage between capital and land assets

At each t, in a risk-free setting, asset markets should function so as to equate the rate of returns to capital and land, otherwise arbitrage can occur. Express the return to assets $r(t) A(t)$ as the return to capital plus total land rent, that is

$$r(t) A(t) = r(t) [K(t) + P_H(t) H] = r(t) K(t) + \Pi(t) H$$

In such a case, a no-arbitrage condition between the two assets is implied. To see this, express the flow budget constraint in terms of assets as

$$\dot{A} = wL + rA - E \tag{4.2}$$

[1] In Chapter 6 the price of capital is endogenous, and the price of land departs from the definition here.

and in terms capital and land as

$$\dot{K} = wL + rK + \Pi H - E \tag{4.3}$$

where expenditure $E = \sum_{j=a,m,s} p_j Q_j$. Therefore

$$\dot{A} - rA = \dot{K} - rK - \Pi H$$

Substituting $A = K + P_H H$ into this equation we get

$$r = \frac{\Pi}{P_H} + \frac{\dot{P}_H}{P_H} \tag{4.4}$$

Consequently, using the flow budget constraint (4.3) presumes an environment which guarantees that the returns to the two assets are equalized at each instant of time.[2] The left hand-side of (4.4) represents the return to the household from one unit of income invested in physical capital. This same unit of income can also buy a quantity of $1/P_H$ of land, generating, at time $t+dt$, a rent income equal to Π/P_H plus the rate of change in the price of land. If this condition did not hold, optimizing investors could exploit the arbitrage opportunity and move investments out of land and into capital. Hence, (4.4) is referred to as the no-arbitrage condition.

4.1.2 Intra-temporal behavior of the household

Household preferences over goods q_a, q_m, and q_s, expressed in per worker terms, $q_j(t) = Q_j(t)/L(t)$, are represented by

$$\int_0^\infty \frac{u\left(q_a(t), q_m(t), q_s(t)\right)^{1-\theta} - 1}{1-\theta} e^{(n-\rho)t} dt \tag{4.5}$$

The felicity function $u(\cdot)$ satisfies Assumption 1.1. In the empirical applications throughout the book, we assume the felicity function is Cobb-Douglas. Preferences for the Cobb-Douglas

[2] In terms of units per effective worker, the result is $r = \hat{\pi}/\hat{p}_H + \dot{\hat{p}}_H/\hat{p}_H + (x+n)$ where $\hat{\pi} = \Pi/\mathcal{A}(t)L(t)$ and $\hat{p}_H = P_H/\mathcal{A}(t)L(t)$.

case are repesented by $u(q_a, q_m, q_s) = q_a^{\lambda_a} q_m^{\lambda_m} q_s^{\lambda_s}$ where λ_j is the share of total expenditure spent on good j, and $\sum_{j=a,m,s} \lambda_j = 1$. As in the two-sector model of the previous chapter, the parameter $\rho > 0$ is rate of time preference. The elasticity of intertemporal substitution is given by $1/\theta$, where $\theta > 0$.

The representative household's flow budget constraint (4.3), expressed in per worker terms, is

$$\dot{k} = w + k(r - n) + \pi H - \epsilon \tag{4.6}$$

where $k = K/L$, $\pi = \Pi/L$ and expenditure per worker $\epsilon = E/L(t)$.[3] Similar to the two-sector model, household expenditure at an instant in time is defined as

$$\mathcal{E}(p_a, p_s) q \equiv$$

$$\min_{q_a, q_m, q_s} \left\{ \sum_{j=a,m,s} p_j q_j \; : \; q \le u(q_a, q_m, q_s), \; (q_a, q_m, q_s) \in \mathbb{R}^3_{++} \right\}$$

where $\mathcal{E}(p_a, p_s) q$ satisfies the properties of $E1-E6$ of Chapter 2, and the price p_m of the manufactured good is the numeraire. The household's problem is to choose $\{q(t)\}_{t\in[0,\infty)}$ to maximize (4.5) subject to initial conditions, $K(0)$, and H, the budget constraint (4.6) in each t, and a limitation on borrowing as given by the transversality condition

$$\lim_{t\to\infty} \left\{ k(t) \cdot \exp\left[-\int_0^t [r(v) - n] \, dv \right] \right\} \ge 0$$

As in the previous chapter, form the present-value Hamiltonian and follow the same procedures to obtain the first order conditions necessary for an interior solution. The result

$$\frac{\dot{q}}{q} = \frac{1}{\theta} \left[r - \rho - \frac{\mathcal{E}_{p_s}(p_a, p_s) p_s}{\mathcal{E}(p_a, p_s)} \frac{\dot{p}_s}{p_s} \right]$$

is virtually identical to the Euler condition of previous model. For the special case where $\theta \to 1$, we obtain

$$\frac{\dot{\epsilon}}{\epsilon} = \frac{\dot{q}}{q} + \lambda_s \frac{\dot{p}_s}{p_s} = r - \rho \tag{4.7}$$

[3] Setting initial period labor to equal unity, implies that $\Pi(t)/L(t) = \Pi(t) e^{-nt}$.

where expenditure share on the home-good is

$$\lambda_s = \mathcal{E}_{p_s}(p_a, p_s)\, p_s / \mathcal{E}(p_a, p_s)$$

The transversality condition places a limit on borrowing and assures that the maximand is bounded,

$$\lim_{t \to \infty} [\xi(t)\, k(t)] = 0$$

where the co-state variable $\xi(t)$ is the present value shadow price of capital.

The production relationships in the following section are expressed in units per effective worker. It is therefore useful to express the key results of this section in the same units. The budget constraint (4.6) in units per effective worker is

$$\dot{\hat{k}} = \hat{w} + \hat{k}(r - x - n) + \hat{\pi} H - \hat{\epsilon} \tag{4.8}$$

where, for example, $\hat{w} = we^{-xt}$. As in the previous chapter, with $\dot{\hat{q}}/\hat{q} = \dot{q}/q - x$, the Euler condition for the general case becomes

$$\frac{\dot{\hat{q}}}{\hat{q}} = \frac{1}{\theta}\left[r - \rho - \theta x - \lambda_s \frac{\dot{p}_s}{p_s}\right] \tag{4.9}$$

and for the case of (4.7)

$$\frac{\dot{\hat{\epsilon}}}{\hat{\epsilon}} = \frac{\dot{\hat{q}}}{\hat{q}} + \lambda_s \frac{\dot{p}_s}{p_s} = r - \rho - x \tag{4.10}$$

4.1.3 Firm behavior

The manufacturing and home-good sectors employ technologies

$$Y_j = \mathcal{F}^j(\mathcal{A}(t) L_j, K_j), \quad j = m, s \tag{4.11}$$

satisfying Assumption 2 of Chapter 2, and expressed in intensive from as

$$\hat{y}_j = \frac{Y_j}{\mathcal{A} L} = f^j\left(l_j, \hat{k}_j\right) \tag{4.12}$$

Here $\hat{y}_j = Y_j e^{-(x+n)t}$ is sector j output per effective worker, l_j is the share of workers employed in the sector, and $\hat{k}_j = K_j / A(t) L$ is the amount of capital stock per effective economy-wide worker employed in sector j. The corresponding cost functions are given by

$$C^j\left(\hat{w}, r^k\right)\hat{y}_j \equiv \min_{l_j, \hat{k}_j} \left\{ l_j \hat{w} + r^k \hat{k}_j \ : \ \hat{y}_j \leq f^j(l_j, \hat{k}_j) \right\}, \ j = m, s$$

where $r^k = r + \delta$. As in the previous chapter, $C^j(\cdot)$ satisfies the properties **C**1 – **C**6 of Chapter 2.

Agricultural production is governed by the technology

$$Y_a = \mathcal{F}^a(\mathcal{A}(t) L_a, K_a, \mathcal{B}(t) H) \tag{4.13}$$

where $\mathcal{F}^a(\cdot)$ satisfies Assumption 2 of Chapter 2. Given the sustainability condition associated with the rate of land augmentation, (4.1), the technology expressed in intensive form is

$$\hat{y}_a = \frac{Y_a}{\mathcal{A}L} = f^a(l_a, \hat{k}_a, H)$$

where $\hat{k}_a = K_a / \mathcal{A}L$. The value-added by agriculture's sector specific resource H is defined as

$$\boldsymbol{\pi}^a(p_a, \hat{w}, r^k)H \equiv \max_{l_a, \hat{k}_a} \left\{ p_a f^a(l_a, \hat{k}_a, H) - \hat{w} l_a - r^k \hat{k}_a \right\} \tag{4.14}$$

and $\pi^a(\cdot)$ satisfies properties $\Pi 1$ – $\Pi 5$ of Chapter 2. Here, $\pi^a(p_a, \hat{w}, r^k)$ is the rental rate per unit of land per effective worker required for the rental market among farmers to clear. As explained in Chapter 2, competition among agricultural firms ensures zero profits for the sector,

$$p_a \hat{y}_a - \hat{w} l_a - r^k \hat{k}_a - \hat{\pi} H = 0$$

Assuming differentiability, by Hotelling's lemma the gradients of $\boldsymbol{\pi}^a(p_a, \hat{w}, r^k)H$ yield – what can be referred to as – the partial equilibrium agricultural supply and derived capital and labor demand per economy-wide effective labor, e.g.,

$$y^a(p_a, \hat{w}, r^k)H = \boldsymbol{\pi}^a_{p_a}(p_a, \hat{w}, r^k)H \tag{4.15}$$

4.1.4 Equilibrium

The equilibrium conditions closely parallel the conditions derived in the previous chapter. We nevertheless repeat the derivations here, and then draw upon these results in the next chapter.

Definition and characterization of equilibrium

Given an initial home-good price, $p_s(0)$, initial resource endowments $\{K(0), L(0), H\}$ and constant world market prices, p_a and p_m, a competitive equilibrium for this economy is a sequence of positive home-good prices and capital stock levels $\left\{p_s(t), \hat{k}(t)\right\}_{t\in[0,\infty)}$, household consumption plans

$$\{\hat{q}_a, \hat{q}_m(t), \hat{q}_s(t)\}_{t\in[0,\infty)}$$

factor rental prices

$$\{\hat{w}(t), r(t), \hat{\pi}(t)\}_{t\in[0,\infty)}$$

for labor, capital and land, and production plans

$$\left\{\hat{y}_a(t), \hat{y}_m(t), \hat{y}_s(t), \hat{k}_a(t), \hat{k}_m(t), \hat{k}_s(t), l_a(t), l_m(t), l_s(t)\right\}_{t\in[0,\infty)}$$

such that at each instant of time t,

1. The representative household solves its utility maximization problem,
2. Firms maximize profits subject to their technologies, yielding zero profits
3. Markets clear for
   (a) commodities

$$\begin{aligned} \hat{y}_m - \hat{q}_m - \dot{\hat{k}} - \hat{k}(x+n+\delta) &\lesseqgtr 0 \\ \hat{y}_s - \hat{q}_s &= 0 \\ \hat{y}_a - \hat{q}_a &\gtreqless 0 \end{aligned} \tag{4.16}$$

(b) labor

$$\sum_{j=a,m,s} l_j = 1$$

(c) capital

$$\sum_{j=a,m,s} \hat{k}_j = \hat{k}$$

4. And the no-arbitrage condition between the capital and land assets holds

$$r = \frac{\boldsymbol{\pi}^a(p_a, \hat{w}, r^k)}{\hat{P}_H} + \frac{\dot{\hat{P}}_H}{\hat{P}_H} + (x+n) \tag{4.17}$$

Equilibrium can be characterized by the following conditions. Given the endogenous sequence $\left\{\hat{k}(t), \hat{\epsilon}(t)\right\}_{t\in[0,\infty)}$, of values, the five-tuple sequence of positive values

$$\left\{\hat{w}(t), r^k(t), \hat{y}_m(t), \hat{y}_s(t), p_s(t)\right\}_{t\in[0,\infty)}$$

satisfies the five intra-temporal conditions for each t:

- zero profits in production of the manufactured and the home-good

$$C^j\left(\hat{w}, r^k\right) = p_j,\ j = m, s \tag{4.18}$$

- labor market clearing

$$\sum_{j=m,s} \frac{\partial}{\partial \hat{w}} C^j\left(\hat{w}, r^k\right) \hat{y}_j - \frac{\partial}{\partial \hat{w}} \boldsymbol{\pi}^a\left(p_a, \hat{w}, r^k\right) H = 1 \tag{4.19}$$

- capital market clearing

$$\sum_{j=m,s} \frac{\partial}{\partial r^k} C^j\left(\hat{w}, r^k\right) \hat{y}_j - \frac{\partial}{\partial r^k} \boldsymbol{\pi}^a\left(p_a, \hat{w}, r^k\right) H = \hat{k} \tag{4.20}$$

- and clearing of the market for the home-good

$$\frac{\partial \mathcal{E}\left(p_a, p_s\right) \hat{q}}{\partial p_s} = \hat{y}_s \tag{4.21}$$

Note the similarity of this intra-temporal characterization to that of the static three-sector model of Chapter 2.

This system (4.18), (4.19), (4.20), and (4.21) can, in principle, be solved to express the endogenous variables $\left\{\hat{w}, r^k, \hat{y}_m, \hat{y}_s, p_s\right\}$ as a function of the exogenous variables (p_a, p_m, H), and the remaining endogenous variables $\left(\hat{k}, \hat{\epsilon}\right)$. Thus, a solution $\left\{\hat{k}^*(t), \hat{\epsilon}^*(t)\right\}_{t\in[0,\infty)}$ is sufficient to find a solution for the remaining variables based upon the intra-temporal conditions.

We next proceed to derive some of the reduced form conditions implied by this system. As in the static model, use the zero profit condition (4.18) to express $\hat{w}$ and r^k as a function of, $p_m = 1$, and p_s. Refer to the result as

$$\hat{w} = W(p_s) \tag{4.22}$$
$$r^k = R(p_s) \tag{4.23}$$

The economy's GDP function can be expressed in effective worker terms using total factor earning

$$\hat{w} + r^k\hat{k} + \boldsymbol{\pi}^a(p_a, \hat{w}, r)H$$

where we use substitute (4.22) and (4.23) for $\hat{w}$ and $\hat{r}$ to obtain

$$G\left(p_s, p_a, \hat{k}, H\right) = W(p_s) + R(p_s)\hat{k} + \boldsymbol{\pi}^a(p_a, W(p_s), R(p_s))H \tag{4.24}$$

Since the factor market clearing conditions (4.19) and (4.20) are linear in $\hat{y}_m$ and $\hat{y}_s$, substitute (4.22) and (4.23) into the these conditions and solve for $\hat{y}_m$ and $\hat{y}_s$ as a function of the endogenous variables p_s, and $\hat{k}$.[4] Express the resulting solution for $\hat{y}_m$ and $\hat{y}_s$ as

$$\tilde{y}^s\left(p_s, \hat{k}\right) \equiv y^s\left(p_a, p_s, \hat{k}, H\right) \tag{4.25}$$

and

$$\tilde{y}^m\left(p_s, \hat{k}\right) \equiv y^m\left(p_a, p_s, \hat{k}, H\right) \tag{4.26}$$

[4] An alternative derivation is to simply derive the supply functions from the respective output price gradient of the GDP function.

In an attempt to decrease notational clutter, we adopt the following convention. A function accented with $\tilde{}$ denotes a function for which all exogenous variables are suppressed, e.g., $\tilde{y}^s\left(p_s, \hat{k}\right)$ suppresses the p_a in $y^s(\cdot)$ as does the expenditure function $\tilde{\mathcal{E}}(p_s) = \tilde{\mathcal{E}}(p_a, p_s)$. As we see later, the arguments of such functions are almost always variables whose levels depend on the solution to a system of differential equations. This notational style is used throughout the text.

The supply function for agriculture is obtained by substituting (4.22) and (4.23) for $\hat{w}$ and r^k in the partial equilibrium supply function (4.15). The supply functions (4.25) and (4.26) are linear in $\hat{k}$ for the same reasons as in the static HOS model of Chapter 2.

The steady-state solution and the equations of motion are derived in the next section.

Inter-temporal equilibrium conditions

As with the two-sector model of Chapter 3, assume an interior solution to the steady-state exists. Then, deriving the steady-state values is virtually identical to that of the two-sector model. The first step is to obtain the steady-state values for r^k, p_s, and $\hat{w}$, and then substitute these values into the budget constraint and solve for $\hat{k}$.

If a steady-state exists, the Euler condition (4.10) implies

$$r^{ss} = \rho + x \tag{4.27}$$

Combining the above expression with (4.23) gives,

$$\rho + x + \delta = R(p_s)$$

Assuming $R(\cdot)$ is invertible, the steady-state home-good price, denoted p_s^{ss}, satisfies

$$p_s^{ss} = R^{-1}(\rho + x + \delta) \tag{4.28}$$

and the effective steady-state wage rate is

$$\hat{w}^{ss} = W(p_s^{ss}) \tag{4.29}$$

Before solving for the steady-state capital stock level, we need to derive a reduced form expression for expenditure $\hat{\epsilon}$ in the budget constraint (4.8). If preferences are homothetic, then the home-good market clearing condition (4.21) is expressed as

$$\hat{\epsilon} = \frac{p_s}{\lambda_s}\tilde{y}^s\left(p_s, \hat{k}\right) \tag{4.30}$$

We can now focus on the budget constraint. Substitute (4.30) for $\hat{\epsilon}$, and (4.22) and (4.23) for $\hat{w}$ and r^k, respectively, into the budget constraint to obtain

$$\dot{\hat{k}} = \tilde{K}\left(p_s, \hat{k}\right) \equiv$$

$$W(p_s) + \hat{k}\left(R(p_s) - \delta - n - x\right) + \tilde{\boldsymbol{\pi}}^a(p_s) - \frac{p_s}{\lambda_s}\tilde{y}^s\left(p_s, \hat{k}\right) \tag{4.31}$$

where for notational convenience

$$\tilde{\boldsymbol{\pi}}^a(p_s) = \boldsymbol{\pi}^a\left(p_a, W(p_s), R(p_s)\right)H$$

Knowing p^{ss} from (4.28), (4.31) is a single equation that is linear in $\hat{k}$. The root k^{ss} satisfying (4.31) for $\dot{\hat{k}} = 0$ and $p_s = p^{ss}$ is the steady-state level of capital stock per effective worker. Knowing the steady-state values $\left(r^{ss}, \hat{w}^{ss}, p_s^{ss}, \hat{k}^{ss}\right)$ permits the calculation of the remaining endogenous variables.

As in the two sector Ramsey model, the next step derives two differential equations in $\hat{k}$ and p_s. Clearly, (4.31) is one of the candidate equations. To derive the differential equation for p_s, differentiate the home-good market clearing condition (4.30) with respect to time to get

$$\dot{\hat{\epsilon}} = \frac{1}{\lambda_s}\left[\left(\tilde{y}^s\left(p_s, \hat{k}\right) + p_s\tilde{y}^s_{p_s}\left(p_s, \hat{k}\right)\right)\dot{p}_s + p_s\tilde{y}^s_{\hat{k}}\left(p_s, \hat{k}\right)\dot{\hat{k}}\right] \tag{4.32}$$

The next step replaces $\dot{\hat{\epsilon}}$ in this expression. To this end, express the Euler condition (4.9) in expenditure terms as

$$\dot{\hat{\epsilon}} = \hat{\epsilon}\frac{1}{\theta}\left[R(p_s) - \theta x - \delta - \rho - \lambda_s(1-\theta)\frac{\dot{p}_s}{p_s}\right] \tag{4.33}$$

The final steps are to substitute the home-good market clearing condition (4.30) for $\hat{\epsilon}$ in (4.33), and then use this result to substitute for $\dot{\hat{\epsilon}}$ in (4.32). The resulting equation is linear in $\dot{p}_s$. Solve for $\dot{p}_s$ to obtain

$$\dot{p}_s = \frac{\left[R\left(p_s\right) - \delta - \rho - \theta x\right] p_s \tilde{y}^s\left(p_s, \hat{k}\right) - \theta p_s \tilde{y}^s_{\hat{k}}\left(p_s, \hat{k}\right)\dot{\hat{k}}}{\theta\left[\tilde{y}^s\left(p_s, \hat{k}\right) + p_s \tilde{y}^s_{p_s}\left(p_s, \hat{k}\right)\right] + \tilde{y}^s\left(p_s, \hat{k}\right)\lambda_s\left(1 - \theta\right)} \tag{4.34}$$

Replacing $\dot{\hat{k}}$ by (4.31) completes the equation. For the case of unitary inter-temporal elasticity of substitution, $\theta \to 1$, we have

$$\dot{p}_s = \frac{\left[R\left(p_s\right) - \delta - \rho - x\right] p_s \tilde{y}^s\left(p_s, \hat{k}\right) - p_s \tilde{y}^s_{\hat{k}}\left(p_s, \hat{k}\right)\dot{\hat{k}}}{\tilde{y}^s\left(p_s, \hat{k}\right) + p_s \tilde{y}^s_{p_s}\left(p_s, \hat{k}\right)} \tag{4.35}$$

This result is virtually identical in structure to the corresponding differential equation of previous chapter.

If a steady-state exists such that $r^k = \rho + \delta + \theta x$, for $\theta > 0$ and $\dot{\hat{k}} = 0$, then (4.34) suggests $\dot{p}_s = 0$. As in the two sector model, these differential equations cannot be solved analytically, and they are autonomous because of the restriction (4.1). Otherwise, t can appear as a separate argument. The time elimination method developed by Mulligan and Sala-i-Martin (1991) is used in the empirical examples in this book to solve the autonomous system, although other methods can be used. The non-autonomous case is considered later in this chapter, and the Brunner and Strulik (2002) method for empirically solving non-autonomous systems is discussed in Chapter 9.

4.1.5 Selected comparative statics

Conditions determining the path of $\hat{w}$, and p_s are the same as in the two-sector closed economy model since Equations (4.22) and (4.23) are identical in structure to the corresponding Equations (3.26) and (3.27) of the previous chapter. Thus, the path of p_s

depends upon the factor intensity of sector s relative to sector m. **Claims 2** and **3** of the previous chapter hold for this model as well.

In the case of output supplies, we again draw upon the homogeneity properties of the supply functions discussed in Chapter 2. Consider the case of agriculture. In non-intensive form, output supply is

$$Y_a = \pi^a_{p_a}\left(p_a, \hat{w}, r^k\right) \mathcal{B}\left(t\right) H$$

and its evolution is given by

$$\frac{\dot{Y}_a}{Y_a} = \varepsilon^{Y_a}_w \frac{\dot{\hat{w}}}{\hat{w}} + \varepsilon^{Y_a}_r \frac{\dot{r}^k}{r^k} + x + n \tag{4.36}$$

where, as in the case of the static model, Equation (2.51), the elasticities $\varepsilon^{Y_a}_w = \left(\partial Y_a / \partial \hat{w}\right)\left(\hat{w}/Y_a\right)$, and $\varepsilon^{Y_a}_r = \left(\partial Y_a / \partial r^k\right)\left(r^k/Y_a\right)$ are negative. In the steady-state, Y_a grows at the rate $\eta = x + n$. For $\hat{k}\left(0\right) < \hat{k}^{ss}$, factor rental rates evolve as $\dot{\hat{w}}/\hat{w} > 0$ and $\dot{r}/r < 0$. Thus, the transition $\dot{Y}_a/Y_a$ depends on the intensity of labor in production relative to capital (i.e., the relative magnitude of the elasticities), and consequently its transition path does not necessarily converge monotonically to its long-run rate of growth.

The case of the other two sectors is most easily seen by appealing to the country's GDP function (4.24) and deriving the supply functions for sectors m and s as the gradient of this function. Expressed in non-intensive form and recognizing that p_a and p_m are constant, we have

$$\frac{\dot{Y}_j}{Y_j} = \varepsilon^{Y_j}_{p_s} \frac{\dot{p}_s}{p_s} + \varepsilon^{Y_j}_{AL}\left(x + n\right) + \varepsilon^{Y_j}_{K} \frac{\dot{K}}{K} + \varepsilon^{Y_j}_{\mathcal{B}H}\left(x + n\right), \; j = m, s \tag{4.37}$$

This equation is similar to (2.50) except for the Harrod rate of technological change parameter x. The factor elasticities $\varepsilon^{Y_j}_{AL}$, $\varepsilon^{Y_j}_{K}$, and $\varepsilon^{Y_j}_{\mathcal{B}H}$ sum to unity. If, for example, sector m is more capital intensive than the other two sectors, it can be shown that $\varepsilon^{Y_m}_{K}$ is positive and

$$\varepsilon^{Y_m}_{AL} < \varepsilon^{Y_m}_{K} > \varepsilon^{Y_m}_{\mathcal{B}H}$$

In the steady-state,

$$\frac{\dot{Y}_j}{Y_j} = x + n, \; j = m, s \tag{4.38}$$

The capital intensive sector, say manufacturing, experiences a positive Rybczynski-like effect whenever $\dot{K}/K > (x+n)$. However, in this case, **Claim 1** (Chapter 3) suggests that $\dot{p}_s/p_s \geq 0$. In the long-run, (4.38) must prevail. If the home-good sector is capital intensive relative to the manufacturing sector, then for $\dot{K}/K > (x+n)$, $\dot{p}_s/p_s \leq 0$.

4.2 Stone-Geary preferences

The assumption that consumer preferences are homothetic is not consistent with the observation that the share of total expenditure allocated to food declines as per capita income grows from some low-subsistence level. Non-homothetic preferences also lead to a distribution of income that differs, over time, from the distribution of expenditures, as per Caselli and Ventura (2000).[5]

Recognizing the implications of non-homothetic preferences to savings behavior, Echevarria (1997) showed the share of resources employed in the non-farm sector increases in the process of growth due to capital deepening that becomes less constrained by the need to meet subsistence consumption requirements. Irz and Roe (2000, 2005) used Stone-Geary preferences to show that in the early stages of development capital can accumulate slowly if (i) agricultural productivity is low, (ii) a majority of the population reside in agriculture, and (iii) consumption is at the subsistence level. Low savings and slow capital accumulation retard the growth of the non-farm sector. Under a three-sector Ramsey model with non-homothetic preferences in agriculture and the informal sector goods, Saracoğlu

[5] The appendix to this chapter extends the Caselli and Ventura framework to the three-sector three factor model considered here.

(2008) shows that as per capita income grows with capital deepening, the share of informal employment in total non-agricultural employment declines over time, as suggested by Turkish data. Another aspect of introducing this structure into the model in the presence of growth in factor productivity is to show how time appears as an argument in the model's differential equations. These equations are non-autonomous which complicates their numerical solution.

4.2.1 Household behavior

The felicity function of a representative household with Stone-Geary preferences is most easily illustrated by

$$q = \psi \left(q_a - \gamma_a\right)^{\lambda_a} q_m^{\lambda_m} q_s^{\lambda_s}$$

where $\gamma_a > 0$ is the subsistence parameter and ψ and λ_j are coefficients. Since all households have the same preference structure and face identical prices, the corresponding expenditure function summed over all households can be expressed as

$$p_a^{\lambda_a} p_m^{\lambda_m} p_s^{\lambda_s} Q + p_a L \gamma_a \tag{4.39}$$

where Q is the linear sum over q. To simplify notation, choose ψ so the scale parameter of (4.39) is unity. Notice the appearance of L in this expression. Expressed in intensive form, the household's budget constraint corresponding to (4.8) becomes

$$\begin{aligned} \dot{\hat{k}} = \hat{w} + \hat{k}(r - x - n) + \boldsymbol{\pi}^a(p_a, \hat{w}, r^k)H- \\ \left[p_a^{\lambda_a} p_m^{\lambda_m} p_s^{\lambda_s} \frac{Q}{\mathcal{A}\left(t\right) L\left(t\right)} + p_a \frac{\gamma_a}{\mathcal{A}\left(t\right)}\right] \end{aligned} \tag{4.40}$$

where $\hat{q} = Q/\mathcal{A}\left(t\right) L\left(t\right)$. We see shortly that the presence of $\mathcal{A}\left(t\right)$ in the denominator of this expression will cause the model's differential equations to be non-autonomous.

The household receives utility from the sequence $\{q\left(t\right)\}_{t\in[0,\infty)}$ expressed as a weighted sum of all future flows of utility

$$\int_0^{\infty} \frac{q\left(t\right)^{1-\theta} - 1}{1-\theta} e^{(n-\rho)t} dt \tag{4.41}$$

and chooses positive values to maximize (4.41) subject to the budget constraint (4.40), the stock of initial assets, and a limitation on borrowing as given by the transversality condition.

Form the present value Hamiltonian and take the first-order conditions to obtain the Euler condition. Express the result in per effective worker units

$$\frac{\dot{\hat{q}}}{\hat{q}} = \frac{1}{\theta}\left(r - \rho - \theta x - \lambda_s \frac{\dot{p}_s}{p_s}\right) \tag{4.42}$$

For the case where $\theta \to 1$, we obtain a result similar to (4.10)

$$\frac{\dot{\hat{q}}}{\hat{q}} + \lambda_s \frac{\dot{p}_s}{p_s} = r - \rho - x \tag{4.43}$$

In this case, however,

$$\frac{\dot{\hat{\epsilon}}}{\hat{\epsilon}} \neq \frac{\dot{\hat{q}}}{\hat{q}} + \lambda_s \frac{\dot{p}_s}{p_s}$$

4.2.2 Equilibrium

The characterization of equilibrium is virtually identical to the previous model, and the Stone-Geary preferences have no effect on the values of the variables in the steady-state since $\lim_{t\to\infty} (\gamma_a / \mathcal{A}(t)) = 0$. As noted above, the differential equations depart from the previous specification, and the empirical method used to solve them must accommodate the non-autonomous feature of these equations.

The steady-state equilibrium

The characterization of equilibrium remains identical to conditions (4.18), (4.19), (4.20), and (4.21). Again, let the manufacturing price serve as numeraire. The home-good market clearing condition is

$$\frac{\partial \hat{\epsilon}}{\partial p_s} = \frac{\lambda_s}{p_s} p_a^{\lambda_a} p_s^{\lambda_s} \hat{q} = \tilde{y}^s\left(p_s, \hat{k}\right)$$

which implies

$$\hat{q} = \frac{p_s \tilde{y}^s \left(p_s, \hat{k} \right)}{\lambda_s p_a^{\lambda_a} p_s^{\lambda_s}} \tag{4.44}$$

where the manufactured good price is numeraire, and we use (4.25) for supply $\hat{y}_s$. Consequently, substituting (4.44) for $\hat{q}$ in the expenditure function we obtain

$$\hat{\epsilon} = \frac{p_s \tilde{y}^s \left(p_s, \hat{k} \right)}{\lambda_s} + p_a \frac{\gamma_a}{\mathcal{A}(t)} \tag{4.45}$$

This equation departs from its counterpart (4.30) in two ways. First, λ_s cannot be interpreted as an expenditure share parameter, and second, the term $p_a \left(\gamma_a / \mathcal{A}(t) \right)$ evolves over time.

If a steady-state exists, the Euler condition (4.42) implies

$$r^{ss} = \rho + \theta x$$

Since the zero profit equations remain unchanged, the factor rental rate equations (4.22) and (4.23) remain unchanged. The steady-state values p_s^{ss}, $\hat{w}^{ss}$, and r^{ss} are derived accordingly. Substituting these steady-state values into the budget constraint, and substituting (4.45) for $\hat{\epsilon}$, yields

$$\hat{w}^{ss} + \hat{k}(r^{ss} - x - n) + \boldsymbol{\pi}^a \left(p_a, \hat{w}^{ss}, r^{ss} \right) H - \frac{p_s^{ss} \tilde{y}^s \left(p_s^{ss}, \hat{k} \right)}{\lambda_s} - p_a \frac{\gamma_a}{\mathcal{A}(t)} \tag{4.46}$$

In the limit, the term $\gamma_a / \mathcal{A}(t)$ approaches zero. Since the steady-state is approached asymptotically as t becomes arbitrarily large, it follows that the steady-state value for $\hat{k}^{ss}$ is identical to the steady-state value obtained using (4.31) where $\dot{\hat{k}} = 0$. The remaining steady-state variables are calculated in the same manner as in the previous model.

While the Stone-Geary structure does not affect the steady-state equilibrium, it does affect the transition to the steady-state. The next section derives the differential equations that determine this transition path.

Inter-temporal equilibrium conditions

Proceeding as in the previous model, substitute Equations (4.22), (4.23), and (4.44) into the budget constraint (4.40) to get the reduced form flow budget constraint

$$\dot{\hat{k}} = \tilde{\mathbf{K}}\left(p_s, \hat{k}, t\right) \equiv W\left(p_s\right) + \hat{k}\left(R\left(p_s\right) \quad x - n - \delta\right) + \tilde{\boldsymbol{\pi}}^a\left(p_s\right) - \left[\frac{p_s \tilde{y}^s\left(p_s, \hat{k}\right)}{\lambda_s} + p_a \frac{\gamma_a}{\mathcal{A}\left(t\right)}\right] \tag{4.47}$$

In contrast to (4.31), note the presence of t which causes this expression to be a non-autonomous differential equation.

To derive the differential equation for p_s, proceed as in the previous model. In this case, however, it proves convenient to focus attention on $\hat{q}$ instead of $\hat{\epsilon}$. Totally differentiate the home-good market clearing condition (4.44) to obtain

$$\lambda_s p_a^{\lambda_a} \dot{\hat{q}} = \left[\left(1 - \lambda_s\right) p_s^{-\lambda_s} \tilde{y}^s\left(p_s, \hat{k}\right) + p_s^{(1-\lambda_s)} \tilde{y}^s_{p_s}\left(p_s, \hat{k}\right)\right] \dot{p}_s + p_s^{(1-\lambda_s)} \tilde{y}^s_{\hat{k}}\left(p_s, \hat{k}\right) \dot{\hat{k}}$$

Next, substitute (4.42) for $\dot{\hat{q}}$. Then, in the resulting expression, substitute the home-good market clearing condition (4.44) for $\hat{q}$ and (4.23) for r. Solve for $\dot{p}_s$ to obtain

$$\dot{p}_s = \frac{\left[R\left(p_s\right) - \delta - \rho - \theta x\right] p_s \tilde{y}^s\left(p_s, \hat{k}\right) - \theta p_s \tilde{y}^s_{\hat{k}}\left(p_s, \hat{k}\right) \dot{\hat{k}}}{\theta\left[\tilde{y}^s\left(p_s, \hat{k}\right) + p_s \tilde{y}^s_{p_s}\left(p_s, \hat{k}\right)\right] + \tilde{y}^s\left(p_s, \hat{k}\right) \lambda_s\left(1 - \theta\right)}$$

Notice that this equation appears identical to that of the previous model, (4.34). However, upon substituting (4.47) for $\dot{\hat{k}}$ we obtain a differential equation in level variables p_s, $\hat{k}$ and t. The term $\gamma_a / \mathcal{A}\left(t\right)$ can be viewed, roughly speaking, as "forcing" households to constrain savings in order to consume. However, since $\gamma_a / \mathcal{A}\left(t\right)$ approaches zero as t becomes arbitrarily large, this effect loses force with time. Thus, the Stone-Geary structure affects the time and rate at which the steady-state is approached by dampening the level of savings in transition.

4.3 A numerical example

The numerical example draws on the same Turkish data used in Chapter 3. In this chapter, however, the industrial-agricultural sector is broken out into two sectors: agriculture and industry. Details on the subsector composition of the agricultural and industrial sectors are discussed in the appendix to Chapter 8.

4.3.1 Parameter estimation

A summary of the model's parameter values for this chapter is given in Table 4.1. The value of the parameters θ, ρ, δ, x, n, and $K(0)$ in Table 3.1, and the service sector labor share, β, remain unchanged. The labor and capital share in agriculture are respectively denoted ϕ_l and ϕ_k. It follows that the share of agricultural production cost accruing to land is equal to $1 - \phi_l - \phi_k = 0.079$. The labor share in agriculture is marginally larger than the labor share in services. Removing agriculture from the industrial-agricultural sector causes α, the labor share parameter of the industrial sector, to decline from 0.481 to 0.436, suggesting the industrial sector is capital intensive. For Cobb-Douglas technologies, these parameters imply factor intensity ratios of labor to capital of 0.77 for the industrial sector, 1.18 for agriculture and 1.13 for the service sector. These parameters values are unchanged in the numerical examples of the remaining chapters.

The expenditure shares on food and industry are denoted λ_a and λ_m, respectively. The estimates of expenditure shares on food and the manufactured good imply the expenditure share on services is about 66 percent. This relatively high share of total expenditures allocated to services is the result of aggregating household and government consumption expenditures. In Turkey, as with many countries, government consumption is dominated by services, which here biases up the household expenditure share on services by an additional 18 percentage points. It is for this reason that government is treated as a separate, albeit benign, agent in extensions to this basic model (see the numerical example of Chapter 6).

Table 4.1 Parameters and initial values

θ	ρ	δ	x	n	α
1.26	0.04	0.04	0.019	0.015	0.436
λ_m	λ_a	β	ϕ_l	ϕ_k	$K(0)$ in 2001 Lira
0.163	0.182	0.529	0.541	0.380	6.2×10^{17}

Source: Author estimates and calculations using WDI and the GTAP data

We expect $\hat{k}(0) < \hat{k}^{ss}$, and, hence, the capital stock to grow at a faster rate than the growth in labor services. As in the empirical example of Chapter 3, capital deepening should trigger Rybczynski-like effects that cause industrial output to grow at a more rapid, albeit declining, rate than other sectors of the economy. The zero profit conditions, Equations (4.22) and (4.23), suggest the service good price should converge to its long-run value from below. This result enables the service sector to better compete for the labor and capital resources that would otherwise flow to agriculture and industry. Since agriculture is the most labor intensive sector, the effect of a rising wage rate on that sector's production costs over time should dominate the corresponding effect of the decreased capital rental rate.

4.3.2 Empirical results

The model predicts a rate of growth in GDP per worker of 3.18 percent in year 2002. By year 2014, the economy reaches the half-way point to the steady-state rate of growth in GDP per worker of about 2.6 percent per annum. By the year 2050, the economy is within 5.2 percent of its long-run rate of growth in GDP per worker of 1.9 percent. Table 4.2 shows the capital stock to GDP ratio is initially about 4, as in the two-sector closed economy model of Chapter 3. The capital to GDP ratio grows to 4.7 by 2031. Capital per worker increases by a factor

Table 4.2 Factor income and total expenditure*

Year	GDP	Capital stock	Wage income	Capital rent	Land rent	Total expenditure	Total Saving to GDP ratio
2001	6506	25896	3202	3210	94	3937	0.39
2006	7584	31799	3707	3788	88	4641	0.39
2011	8714	37964	4242	4386	86	5380	0.38
2016	9907	44416	4811	5011	86	6159	0.38
2021	11175	51198	5417	5671	88	6986	0.37
2026	12531	58365	6067	6374	91	7867	0.37
2031	13990	65984	6767	7127	96	8812	0.37

*millions of 2001 Turkish Lira per worker
Source: model results

of 1.47 over the 2001–2011 interval, and by a factor of 2.55 by 2031. At the half-way point to long-run equilibrium, GDP per worker increases by 41 percent, and by a factor of 2.15 by 2031. About 27 years are required for income per worker to double. At this level of analysis, the model predicts outcomes that are only slightly higher than the predictions of the two-sector closed economy model.

Based on the entries in Table 4.2, initially, the share of labor and capital income in total income is equal to about 49.2 percent for labor income, 49.4 for capital, while the share of income from land is 1.4 percent of total income. By 2031, labor's share in total income declines to 48.3 percent, capital increases to 51 percent and land falls to about 0.7 percent. The non-monotonic trajectory of total land rent per worker reflects the differential effects over time of the increase in the wage rate, and the decrease in the capital rental rate. Since agriculture is relatively labor intensive, the rise in the wage rate increases agricultural costs of production at a faster rate than the decline in the capital rental rate. These opposing effects abate as the rate of change in the wage rate slows while the capital rental rate continues to fall.

Table 4.3 shows the evolution of sector value shares in GDP and the share of labor and capital employed in each sector. In

Table 4.3 Sector value shares in GDP and sector factor shares in total factors

	Sector Share in GDP			Labor Share in			Capital Share in		
Year	Industry	Agriculture	Service	Industry	Agriculture	Service	Industry	Agriculture	Service
2001	0.420	0.181	0.399	0.372	0.199	0.429	0.480	0.140	0.380
2006	0.451	0.146	0.403	0.402	0.161	0.437	0.509	0.111	0.380
2011	0.470	0.123	0.407	0.421	0.137	0.442	0.526	0.093	0.380
2016	0.482	0.108	0.410	0.433	0.121	0.447	0.537	0.081	0.381
2021	0.490	0.098	0.412	0.441	0.110	0.450	0.544	0.074	0.382
2026	0.495	0.091	0.414	0.446	0.102	0.452	0.549	0.068	0.383
2031	0.499	0.086	0.415	0.450	0.096	0.454	0.553	0.064	0.383

Source: Model results

contrast to the closed economy model of the previous chapter, sectoral changes are more pronounced. Industrial value share in GDP rises from 42 percent initially to 50 percent by 2031 while the service sector value share rises from 40 percent to only 41.5 percent of GDP. Agriculture's value share in GDP declines from 18 percent in 2001 to 9.6 percent by 2031. These changes correspond to a reallocation of the economy's resources out of agriculture and into the industrial and service sectors of the economy. Most pronounced is the increased share of the labor force in the industrial sector and the labor share in agriculture declining to under 10 percent from a high of 20 percent. The share of total capital employed in the industrial sector increase by 15 percent and remains virtually unchanged in the service sector. Capital deepening occurs in all sectors. In the case of agriculture, the capital to labor ratio increases by a factor of 1.2 between 2001 and 2006, and by a factor of 2.4 by 2031. While Table 4.2 shows a non-monotonic pattern for the evolution of land rental income per worker, when measured as land rental income per agricultural worker, it rises by a factor of 1.15 over the period 2001–2006, and by 2.1 over the 2001–2031 period.

Tables 4.4, 4.5, and 4.6 show the effects of capital deepening on changes in sectoral contributions to GDP, and to changes in input share. These tables are the result of a growth accounting exercise using Equations (4.36) and (4.37). The dominant features revealed in Table 4.4 are the positive Rybczynski-like effects of growth in the capital stock on industrial sector output growth, and growth in labor services on service sector output growth. The growth in output of both these sectors converge from above to their long-run rate of growth per worker of 1.9 percent per annum. The basic economic forces unveiled in the growth accounting exercise are similar to those discussed in the numerical example of Chapter 3. As capital deepening occurs, the industrial sector experiences a growth in output as the decline in the capital rental rate lowers the largest contributor to its total cost of production. As the sector employs a larger share of the economy's total capital services, the sector's productivity of labor rises. Since growth in capital stock converges from above

Table 4.4 Growth in industrial output and factor contributions

Year	Growth in output	Contributions to Growth: Value added price	Capital stock	Effective labor
2001	0.0660	−0.1383	0.3949	−0.1905
2006	0.0542	−0.0985	0.3304	−0.1777
2011	0.0475	−0.0724	0.2908	−0.1708
2016	0.0434	−0.0543	0.2644	−0.1667
2021	0.0407	−0.0412	0.2460	−0.1641
2026	0.0389	−0.0314	0.2327	−0.1624
2031	0.0375	−0.0242	0.2229	−0.1612

Source: Model results

to the growth in labor services, if all else were constant, capital deepening in the industrial sector would increase its marginal value product of labor relative to the growth in the marginal value product of labor in the service sector.

Given homothetic preferences, service sector demand increases as household income increases – see Table 4.2. Accordingly, service sector firms need to employ proportionately more labor than the industrial sector to meet the increased demand. While capital deepening contributes to a decline in the unit cost of service sector production, the relatively smaller share of capital in the service sector's total cost is not a sufficient incentive for service sector firms to increase output to meet final demand. Consequently, service good market clearing responds with an increase in the service good price. The price of the service good in terms of the numeraire price of the industrial good increases by 3 percent over the period of transition. This increase contributes 12 percentage points to service sector output in 2006, and declines as the rate of capital deepening approaches the rate of growth in labor services (see Table 4.5).

Agriculture is the most labor intensive sector in the economy. Without factor augmentation to land, i.e., the sustainability condition (4.1), agricultural output per *effective* worker would converge to zero in the long-run if $\eta < x + n$. The rise in the

Table 4.5 Growth in service output and factor contribution

		Contributions to Growth		
Year	Growth in output	Value added price	Capital stock	Effective labor
2001	0.0476	0.1621	−0.3427	0.2282
2006	0.0442	0.1203	−0.3045	0.2284
2011	0.0417	0.0902	−0.2767	0.2281
2016	0.0398	0.0683	−0.2562	0.2277
2021	0.0383	0.0520	−0.2410	0.2273
2026	0.0373	0.0398	−0.2295	0.2269
2031	0.0364	0.0306	−0.2208	0.2266

Source: Model results

wage rate increases agriculture's cost of production in greater proportion than the decline in cost accompanying the fall in the capital rental rate. The wage effect dominates the positive interest rate effect throughout transition (Table 4.6). As both the wage effect and interest rate effect decline, the growth in agriculture's output converges from below to that of the other sectors in the long-run due to the exogenous technological change in land services. As labor departs agriculture, agriculture's capital to labor ratio rises as does its capital to output ratio. This ratio ranges from 3.1 in the initial period to 3.6 in the long-run.

Table 4.6 Growth in agricultural output and factor contributions

		Contributions to Growth		
Year	Growth in output	Wage effect	Interest rate effect	Technical change
2001	−0.0026	−0.0792	0.0431	0.0336
2006	0.0057	−0.0612	0.0333	0.0336
2011	0.0120	−0.0473	0.0257	0.0336
2016	0.0170	−0.0365	0.0198	0.0336
2021	0.0207	−0.0282	0.0153	0.0336
2026	0.0236	−0.0218	0.0119	0.0336
2031	0.0259	−0.0169	0.0092	0.0336

Source: Model results

However, agriculture's total capital stock remains virtually unchanged throughout transition.

The effect of these differential rates of sector growth cause the economy to experience a trade reversal. As a small open economy facing given and fixed industrial and agricultural good border prices, the economy initially exports the agricultural good and imports the industrial good. The value of agricultural export plus the value of industrial imports account for over 14 percent of GDP initially, and then declines to zero in about 2014, after which the economy exports the industrial good and imports the agricultural good. In the long-run, the value of trade accounts for about 8 percent of GDP.

4.4 Conclusion

This chapter drew upon the static three-sector model developed in Chapter 2 and the two-sector Ramsey model presented in Chapter 3, to develop a three-sector model of a small open economy in which one of the three sectors employs a sector specific factor. The derivation of the model's intra-temporal and inter-temporal equilibrium conditions are shown to follow a pattern similar to that of the simpler two sector closed economy model. The model is shown to have the same steady-state properties for the case of homothetic and Stone-Geary preferences. In the absence of growth in factor productivity, the sector specific factor allows for diminishing returns to labor and capital to occur more rapidly than in the other sectors, while the evolution of the price of the home-good relative to traded goods explains the process by which the non-traded sector competes with the traded good sectors for resources. The asset market also received attention because capital and the sector specific factor are the two assets held by households. In the process of growth, the decline in the rental rate of capital and the change in the rental rate of the sector specific factor imply an evolution of the price of the sector specific factor relative to the price of capital. This issue is taken up in the next chapter.

The numerical example drew upon the same data as the example presented in Chapter 3. The results predict a structural transformation of the Turkish economy that entails the pulling of labor from agriculture as capital deepening occurs, and changes in the shares of labor and capital employed in the three sectors. The service sector is treated as the numerical counterpart of the model's home-good sector, and plays a key role in competing for resources with the two traded good sectors of the economy. In spite of these changes, the growth in GDP, factor income from labor and capital, and the rate of transition to long-run equilibrium are not substantially different from that generated by the two-sector, closed economy model.

The three-sector model serves as a point of departure for the models presented in the remaining chapters of the book, including the two-country world model presented in Chapter 7. This basic model also serves as foundation upon which other sectors can be added to the framework. Including agriculture as the sector employing a sector specific factor helps to focus on the supply of surplus labor from the farm sector to the rest of the economy as an essential part of the growth process. Other applications may focus on the factor specificity of environmental services or energy, as in Gaitan and Roe (2005). The next chapter extends this basic three-sector model to account for intermediate factors of production, composite capital and government.

4.5 Appendix: income and expenditure distribution

This section builds upon the work of Caselli and Ventura (2000). Knowing the solution to the Ramsey model's representative agent problem, they show how this solution can be used to determine the ith agent's income and expenditure level at each instant in time relative to that of the model's representative agent. This section shows how their contribution applies to the model presented in this chapter.[6]

---

[6]We express appreciation to Harumi Nelson for contributing this material. See Nelson et al. (2009) for an empirical application of this approach.

The economy consists of a large number of households, each of which is of equal family size $L(t)$ and is an infinitely lived dynasty. Household i $(i = 1, 2, ..., I)$ maximizes utility which takes the same form as (4.5). The felicity function is Stone-Geary form, given by $q_i = u(q_{ai}, q_{mi}, q_{si}) = (q_{ai} - \gamma)^{\lambda_a} q_{mi}^{\lambda_m} q_{si}^{\lambda_s}$ where q_{ji} is household i's per worker level of consumption of good j. All parameters are identical across the households in the economy. The flow budget constraint for household i is given by the same form as Equation (4.6):

$$\dot{k}_i = wl_i + k_i (r - n) + \pi H_i - \epsilon_i \tag{4.48}$$

where k_i and ϵ_i are household i's per worker capital asset and expenditure, respectively. We assume that labor, l_i, and land, H_i, are given exogenously and normalized such that $(1/I)\sum_i l_i = 1$, and $(1/I)\sum_i H_i = 1$. The per worker level of total spending in household i is given by

$$\epsilon_i = p_a \gamma + p_a^{\lambda_a} p_s^{\lambda_s} q_i \tag{4.49}$$

For our purposes here, per worker expenditure is separated into two components, (i) the amount spent on the agricultural good required to meet subsistence needs, $p_a \gamma$, and (ii) the amount spent of the agricultural good in excess of basic needs plus the remaining consumption goods, $p_a^{\lambda_a} p_s^{\lambda_s} q_i$. We refer to the former as subsistence expenditure which is identical for all households, and the latter as the supernumerary expenditure which we denote by $\mu_i = p_a^{\lambda_a} p_s^{\lambda_s} q_i$. Substituting (4.49) into (4.48), and solving the usual Hamiltonian problem, we obtain the Euler condition as

$$\frac{\dot{\mu}_i}{\mu_i} = \frac{1}{\theta}\left[(r - \rho) - (1 - \theta)\lambda_s \frac{\dot{p}_s}{p_s}\right] \tag{4.50}$$

The transversality condition for the representative household is applied for each household i. Using Equations (4.48), (4.50) and the transversality condition, we obtain household i's (supernumerary) expenditure function

$$\mu_i = \zeta \omega_i = \zeta(k_i + \omega_w l_i + \omega_\pi H_i - \omega_\gamma) \tag{4.51}$$

where the supernumerary expenditure is a fraction, ζ (the propensity to consume out of wealth) of total wealth which consists of the net asset holdings, the present values of income from wages, land rent, and the present value of subsistence expenditure.[7]

The distribution dynamics are discussed in terms of the individual household's relative position to the corresponding mean. The relevant mean values are defined simply as $\epsilon = (1/I)\sum_i \epsilon_i$, $q = (1/I)\sum_i q_i$, $\mu = (1/I)\sum_i \mu_i$, and $k = (1/I)\sum_i k_i$. Since the propensity to consume out of wealth is independent of individual characteristics, we obtain the economy's expenditure function by aggregating (4.51) over households. The resulting expenditure function for the representative household is expressed as (4.51) above, omitting the subscript for a household. Similarly, the budget and the Euler condition for the representative household take the same form as (4.48) and (4.50), respectively, except for household index i. Thus, we observe the average variables over all households in this economy as the behavior of the representative household in the Ramsey framework.

4.5.1 Distinguishing individual expenditure from that of the representative household

The growth rate of expenditure is large when a high share of expenditure is attached to the supernumerary spending which, over time, changes by the same proportion as other households' in the economy. The evolution of household i's per worker expenditure relative to the representative household is shown by

---

[7] The respective definitions are given by:

$$\zeta(t)^{-1} = \int_t^{\infty} \exp\left[-(\rho/\theta - n)(\tau - t) + \frac{1-\theta}{\theta}\int_t^{\tau}(r(v) - \lambda_s \frac{\dot{p}_s(v)}{p_s(v)})dv\right] d\tau,$$

$$\omega_z(t) = \int_t^{\infty} \exp\left[-\int_t^{\tau} [r(v) - n]\, dv\right] z(\tau)d\tau$$

for $z = w, \pi$ and

$$\omega_\gamma(t) = p_a \gamma \int_t^{\infty} \exp\left[-\int_t^{\tau} [r(v) - n]\, dv\right] d\tau.$$

$$\frac{d(\epsilon_i/\epsilon)}{dt} = \dot{\phi}(1 - \frac{\epsilon_i}{\epsilon}) \tag{4.52}$$

where $\dot{\phi} = \frac{\dot{\epsilon}}{\epsilon} - \frac{\dot{\mu}}{\mu}$.[8] The sign of $\dot{\phi}$ is negative when the mean expenditure increases as the economy converges to its long-run equilibrium from a level of initial capital stock that is less than its long-run equilibrium level. Otherwise, $\dot{\phi}$ is positive.

Therefore, in the case where the mean expenditure is increasing, $\dot{\phi} < 0$, households who initially spend less (more) relative to the mean will spend a smaller (larger) proportion relative to the mean household in the later periods. In other words, the poor's expenditure (i.e., those whose expenditures are less than the representative household) does not grow as fast as the representative household over time. In this case, the poor's relative expenditure position deteriorates in the later periods compared to the representative household. The reverse argument holds when the value of $\dot{\phi}$ is positive. At the steady-state, the value is zero and there is no evolution in the relative per worker expenditure level.

4.5.2 Distinguishing individual income from that of the representative household

Expenditure function (4.51) shows a household that is endowed with a small share of the capital asset tends to improve (worsen) its relative asset position over time when the other sources of household wealth grow slower (faster) than total wealth.[9] To

---

[8] To see this result, note that

$$\begin{aligned} \frac{d(\epsilon_i/\epsilon)}{dt} &= \frac{1}{\epsilon}\left(\dot{\mu}_i - \dot{\mu}\frac{\epsilon_i}{\epsilon}\right) \\ &= \frac{1}{\epsilon}\frac{\dot{\mu}}{\mu}\frac{(p_a\gamma + \mu)\,\mu_i - \mu\,(p_a\gamma + \mu_i)}{\epsilon} \\ &= \left(\frac{\dot{\epsilon}}{\epsilon} - \frac{\dot{\mu}}{\mu}\right)\left(1 - \frac{\epsilon_i}{\epsilon}\right) \end{aligned}$$

Note also that there is no inter-household dynamics in the expenditure distribution if $\gamma = 0$.

[9] Total wealth equals $\omega_i = \frac{\mu_i}{\mu}/\frac{\xi}{\mu}$, the wealth share of wage equals $\frac{\omega_w l_i}{\omega_i} = \frac{\omega_w}{\mu/\xi}\frac{l_i}{\mu_{i/}/\mu}$ and that of land rent equals $\frac{\omega_\pi H_i}{\omega_i} = \frac{\omega_\pi}{\mu/\xi}\frac{H_i}{\mu_{i/}/\mu}$. Thus, a small (large) share of the capital

see the linkage between the share of each wealth component and the growth rate of capital assets more rigorously, consider the differential equation for the relative asset position of household i at period t:

$$\frac{d(k_i/k)}{dt} = \Omega_w(l_i - \frac{k_i}{k}) + \Omega_\pi(H_i - \frac{k_i}{k}) - \Omega_\gamma(1 - \frac{k_i}{k}) \qquad (4.53)$$

where $\Omega_w = (w - \zeta\omega_w)/k$, $\Omega_\pi = (\pi - \zeta\omega_\pi)/k$ and $\Omega_\gamma = (p_a\gamma - \zeta\omega_\gamma)/k$. These terms are interpreted as the net savings out of: wage income, rent income from H_i, and subsistence expenditure, all as the capital asset share, respectively, for the representative household.[10] Equation (4.53) indicates that large relative labor skill as well as land ownership are likely to contribute to the improvement (deterioration) of the relative position capital asset when the net savings out of wage income and rent income from land are positive (negative).

Consider a special case in which there is no heterogeneity in labor skill and land ownership with $\theta = 1$. In this case, the dynamics of the distribution of assets is characterized by the difference in the average growth rates between the per worker capital asset and the supernumerary expenditure.[11] Suppose the interest rate is greater than the time preference rate so that the growth rate of the supernumerary expenditure is positive from (4.50). Then, the capital asset accumulates faster than the supernumerary expenditure and as a result, the relative asset position of a low asset holder improves while the relative position of a large asset holder worsens over time. Once the heterogeneity is extended to labor skill as well as land ownership, the evolution of capital asset depends on which component outweighs other elements in (4.53).

---

asset is a likely result of large (small) labor endowment and land ownership compared to (supernumerary) expenditure.

[10] Integrating (4.53), we obtain

$$\begin{aligned} \frac{k_i(t)}{k(t)} &= \frac{k_i(0)}{k(0)} e^{-\int_0^t (\Omega_w(s)+\Omega_\pi(s)-\Omega_\gamma(s))ds} \\ &\quad + \int_0^t (\Omega_w(\tau)l_i + \Omega_\pi(\tau)H_i - \Omega_\gamma(\tau)) e^{\int_t^\tau (\Omega_w(s)+\Omega_\pi(s)-\Omega_\gamma(s))ds} d\tau \end{aligned}$$

[11] That is, $\frac{d(k_i/k)}{dt} = (\frac{\dot{k}}{k} - \frac{\dot{\mu}}{\mu})(1 - \frac{k_i}{k})$.

Define household i's per worker level of income by $m_i = wl_i + rk_i + \pi H_i$. Then, the relative position of income to the mean is given by

$$\frac{m_i}{m} = s_l l_i + s_H H_i + s_k \frac{k_i}{k} \tag{4.54}$$

where $s_l = w/m$, $s_k = rk/m$ and $s_H = \pi/m$ are the respective labor income share, capital income share and land income share of the representative household. The position of the relative income is obtained by substituting the solution for the relative asset position from (4.53) into Equation (4.54).[12]

In order to explain what economic variables contribute to changes in the relative position of income, we differentiate the equation for the relative income position above and express it as follows

$$\frac{\dot{m}_i}{m_i} - \frac{\dot{m}}{m} = \frac{\dot{w}}{w}\left(\frac{l_i}{m_i/m} - 1\right) s_l + \frac{\dot{\pi}}{\pi}\left(\frac{H_i}{m_i/m} - 1\right) s_H +$$

$$\left(\frac{\dot{r}}{r} + \frac{\dot{k}}{k}\right)\left(\frac{k_i/k}{m_i/m} - 1\right) s_k + \frac{k_i/k}{m_i/m}\left(\frac{\dot{k}_i}{k_i} - \frac{\dot{k}}{k}\right) s_k \tag{4.55}$$

The difference in proportional changes of income between household i and the representative household can be decomposed into four components: (direct) contributions from labor and land as well as capital asset holdings, and an additional effect from the proportional change in the capital asset holdings relative to the representative household. Each component of the direct contributions from the factors of income is expressed as the growth rate of earnings multiplied by its weight.[13] Since the weights that are attached to the growth rate of earnings sum to zero, a part or all of the direct effects from wage and land rent is

---

[12] This equation suggests the dynamics of the relative income position is largely influenced by the evolution of the relative asset holdings when the income shares of each source is relatively constant over time.

[13] Each of these weights, $((\frac{l_i}{m_i/m} - 1)s_l$, $(\frac{H_i}{m_i/m} - 1)s_H$, and $(\frac{k_i/k}{m_i/m} - 1)s_k)$ measures the deviation of individual household's factor income share from the representative household. For instance, an increase in wage has a positive effect on the improvement of household i's income position when the household labor income share is larger than the average labor income share. A similar argument holds for the relationship between the relative income position and land ownership.

cancelled by the direct effect of the capital asset. Thus, the evolution of the relative asset position, which is shown as the last term in (4.55), is the most important component determining the relative income position of the ith household. As a special case where there is no heterogeneity in labor and land with $\theta = 1$, the growth rate of relative income is determined by a simple form.[14] In this case, when an economy's stock of capital per worker is increasing, the poor's relative position of income improves, relative to the mean, and the high income household's position, relative to the mean, worsens along with its asset position.

[14] It is given by $\frac{d(m_i/m)}{dt} = (\frac{\dot{m}}{m} - \frac{\dot{r}}{r} - \frac{\dot{\mu}}{\mu})(1 - \frac{m_i}{m})$.

5

Extensions to the Three-Sector Model

Chapters 3 and 4 present the theoretical foundations of multi-sector closed and open economy models. These models are useful for understanding the basic forces of economic growth, but are highly stylized. The purpose of this chapter is to extend the basic models in three ways: introduce (i) intermediate inputs, (ii) composite capital, and (iii) government. We introduce separately each of these extensions for pedagogical reasons which also keeps the notation relatively simple, and for parsimony which is often desirable in empirical applications.

We first discuss intermediate inputs, a feature commonly found in the static general equilibrium literature. Adding this feature to the model provides insights into how the growth process can influence a country's domestic terms of trade. A model with intermediate inputs also provides the analyst with a tool to examine the economics of vertical market structures, e.g., primary agriculture and the food processing industry, or crude oil and the refinery industry. Next, we model capital stock as a composite of the output of all sectors of the economy. This modification is important because data show that a country's capital stock is more than bricks, mortar and machinery; it also includes software and various products from agriculture. A model with composite capital still has a single state variable, but since the home-good price is endogenous, it follows that the price of capital becomes endogenous. In this environment, household asset values depend on both changes in the stock of capital and changes in the underlying price of capital. In spite of this complication, the numerical algorithms used to solve such a problem still rely on the solution to a system of two autonomous differential equations. The chapter concludes with a model of government consumption and taxes. This modification is essential

T.L. Roe et al., *Multisector Growth Models*, DOI 10.1007/978-0-387-77358-2_5,

for policy analyses concerned with the economic growth effects of indirect taxes, barriers to trade, government expenditures in excess of fiscal revenues and taxes on household income.

The next chapter combines each of these extensions in a single model that can serve as a more general purpose policy analysis tool. This model is made more complex by the linkages between these extensions, but one that can still be specified in a manner common to the specifications in this chapter.

5.1 Intermediate inputs of production

A unit of land, labor, or capital is a stock variable that provides a flow of services each period, and subject to depreciation, is reusable. An intermediate input is a flow resource that depreciates fully once used. Country input-output tables tend to show a sector's own output accounts for the largest share of its value of total output compared to other intermediate inputs. For many countries, and depending on the level of aggregation, the service sector's output tends to account for the next highest share of intermediate cost in value of total output. Examples of important intermediate service sector activities in this category are computer and professional services, financial services including insurance, and utilities, construction and logistical services of various kinds.[1] For many countries, these sectors tend to be isolated from world markets making the employment of best practice technologies problematic.

The motivation for including intermediate inputs is therefore several. Hirschman (1958) emphasized the importance of intermediate goods in sectoral linkages to economic development, and Yi (2003) observes that tariffs can multiply up when goods are traded multiple times during stages of production. Jones (2007) introduces intermediate inputs into a Solow economy. He shows the multiplier effects of a shock to productivity far exceed the same shock without intermediates. Intermediate inputs also link

[1] See Bhagwati et al. (2004) for a discussion of the dichotomy of trade in services.

traded and home-goods. An increase in service sector productivity, all else constant, implies that fewer resources are needed to satisfy intermediate demand for services in the traded good sectors of the economy. In this case, the traded good sectors face a less negative domestic terms of trade effect in the process of growth: in other words, a positive productivity shock to the service sector can release resources to the traded good sectors of the economy.

5.1.1 Firms

Relative to the three-sector model of Chapter 4, the household side of the economy remains unchanged when introducing intermediate inputs. We therefore begin with the specification of firm technologies.

Firms in manufacturing and home-good production employ a technology with the following structure:

$$Y_j \leq \min_{L_j, K_j, Y_{aj}, Y_{mj}, Y_{sj}} \left\{ \mathcal{F}^j \left(\mathcal{A}(t) L_j, K_j \right), \frac{Y_{aj}}{\sigma_{aj}}, \frac{Y_{mj}}{\sigma_{mj}}, \frac{Y_{sj}}{\sigma_{sj}} \right\}, \; j = m, s \tag{5.1}$$

Here $\mathcal{F}^j(\cdot)$ satisfies Assumption 2.2 in the primary factors labor and capital, and Y_{aj}, Y_{mj} and Y_{sj} denote the amount of the manufacturing, agricultural and home-good employed as intermediate inputs in producing the manufacturing and home-good respectively. The σ_{ij} are input-output coefficients that determine the amount of intermediate factor Y_{ij} required to produce one unit of output Y_j. In the case of agriculture, the technology is

$$Y_a \leq \min_{L_a, K_a, Y_{aa}, Y_{ma}, Y_{sa}} \left\{ \mathcal{F}^a(\mathcal{A}(t) L_a, K_a, \mathcal{B}(t) H), \frac{Y_{aa}}{\sigma_{aa}}, \frac{Y_{ma}}{\sigma_{ma}}, \frac{Y_{sa}}{\sigma_{sa}} \right\} \tag{5.2}$$

where $\mathcal{F}^a(\cdot)$ satisfies Assumption 2.2 in the primary factors labor, capital and land.

As in the previous model, we normalize each technology by the economy-wide effective labor force $\mathcal{A}(t) L(t)$. Then the corresponding minimum total cost of producing $\hat{y}_j$ units of output is

$$\left(C^j\left(\hat{w}, r^k\right) + \sum_{i=a,m,s} \sigma_{ij} p_i\right) \hat{y}_j, \; j = m, s \tag{5.3}$$

The product $\sigma_{ij} p_i$ is the cost of the intermediate input $\hat{y}_{ij}$ per unit of output $\hat{y}_j$ per effective worker. We remind the reader that $\hat{w}$ is the effective wage rate, and r^k is the rate paid by firms to employ the services of capital.

Agricultural firms rent land from households. Competition among firms implies profits equal returns to the sectors specific factor, land. Consequently, as in the previous model, it is convenient to maximize firm profits subject to (5.2). The result leads to the restricted value-added function per economy-wide effective worker

$$\boldsymbol{\pi}^a\left(p_{va}, \hat{w}, r^k\right) H \tag{5.4}$$

which satisfies properties Π1–Π5 of Chapter 2. In (5.4), value-added price of the agricultural good is $p_{va}(t) = p_a - \sigma_{aa} p_a - \sigma_{ma} p_m - \sigma_{sa} p_s(t)$.

5.1.2 *Equilibrium with intermediate inputs of production*

We dispense with the definition of equilibrium since it remains virtually unchanged from the previous model, and proceed with the characterization of equilibrium.

Intra-temporal equilibrium

For each t, and corresponding capital stock and expenditure pair $\left\{\hat{k}(t), \hat{e}(t)\right\}_{t\in[0,\infty)}$, an intra-temporal equilibrium is characterized by the five-tuple sequence of positive values

$$\left\{\hat{w}(t), r^k(t), \hat{y}_m(t), \hat{y}_s(t), p_s(t)\right\}_{t\in[0,\infty)}$$

that satisfy the following five equations:

- zero profit in manufacturing and services

$$C^j\left(\hat{w}, r^k\right) = p_{vj}, \quad j = m, s \tag{5.5}$$

- labor market clearing

$$\sum_{j=m,s} C^j_{\hat{w}} \left(\hat{w}, r^k\right) \hat{y}_j - \boldsymbol{\pi}^a_{\hat{w}} \left(p_{va}, \hat{w}, r^k\right) H = 1 \tag{5.6}$$

- capital market clearing

$$\sum_{j=m,s} C^j_{r^k} \left(\hat{w}, r^k\right) \hat{y}_j - \boldsymbol{\pi}^a_{r^k} \left(p_{va}, \hat{w}, r^k\right) H = \hat{k} \tag{5.7}$$

- and home-good market clearing

$$\frac{\partial \mathcal{E}\left(p_a, p_s\right) \hat{q}}{\partial p_s} = \hat{y}_s \left(1 - \sigma_{ss}\right) - \sigma_{sm} \hat{y}_m - \sigma_{sa} \hat{y}_a \tag{5.8}$$

- The value-added prices in (5.5) are given by

$$\begin{aligned} p_{vm} &= p_m \left(1 - \sigma_{mm}\right) - \sigma_{am} p_a - \sigma_{sm} p_s \\ p_{vs} &= p_s \left(1 - \sigma_{ss}\right) - \sigma_{as} p_a - \sigma_{ms} p_m \end{aligned}$$

The main difference between the characterization here, and that of the equilibrium in Chapter 4 is the presence of value-added prices, p_{vj} and additional terms in the home-good market clearing condition (5.8). These terms account for intermediate demand of the home-good in own production and in production of the agricultural and the manufacturing good. As in the previous model, this system of equations can, in principle, be solved to express the endogenous variables $\left(\hat{w}, r^k, \hat{y}_m, \hat{y}_s, p_s\right)$ as a function of the exogenous variables $\left(p_m, p_a, H\right)$, and the remaining endogenous variables $\left(\hat{k}, \hat{\epsilon}\right)$. The equilibrium solution $\left\{\hat{k}^*\left(t\right), \hat{\epsilon}^*\left(t\right)\right\}_{t\in[0,\infty)}$ is then sufficient to find a solution to the remaining variables using the intra-temporal conditions.

Use the zero profit conditions (5.5) to express r^k and $\hat{w}$ as a function of value-added prices p_{vm}, and p_{vs}, both of which now depend on the endogenous price p_s. We express this result as

$$\tilde{w}\left(p_s\right) \equiv W\left(p_{vm}, p_{vs}\right) \tag{5.9}$$

and

$$\tilde{r}(p_s) \equiv R(p_{vm}, p_{vs}) \tag{5.10}$$

These functions are homogeneous of degree one in value-added prices p_{vm} and p_{vs}, and thus convey the same Stopler-Samuelson like properties of the static three-sector model.

Substitute (5.9) and (5.10) into the factor market clearing conditions, which are linear in $\hat{y}_m$ and $\hat{y}_s$, to obtain the supply functions

$$\hat{y}_m = \tilde{y}^m(p_s, \hat{k}) \equiv y^m(p_{vm}, p_{va}, p_{vs}, \hat{k}, H) \tag{5.11}$$

and

$$\hat{y}_s = \tilde{y}^s(p_s, \hat{k}) \equiv y^s(p_{vm}, p_{va}, p_{vs}, \hat{k}, H) \tag{5.12}$$

The supply function for agriculture is given by

$$\hat{y}_a \equiv \boldsymbol{\pi}^a_{p_{va}}\left(p_{va}, \hat{w}, r^k\right) H$$

Substituting (5.9) and (5.10) for $\hat{w}$ and r^k into this equation yields a reduced form supply function for agriculture:

$$\hat{y}_a = \tilde{y}^a(p_s) H \equiv \boldsymbol{\pi}^a_{p_{va}}\left(p_{va}, W(p_{vm}, p_{vs}), R(p_{vm}, p_{vs})\right) H \tag{5.13}$$

For later reference, denote the value-added by land H as

$$\tilde{\pi}(p_s)H \equiv \boldsymbol{\pi}^a\left(p_{va}, \tilde{w}(p_s), \tilde{r}(p_s)\right) H \tag{5.14}$$

Inter-temporal equilibrium

Since the household's problem remains unchanged from the previous model, we use the Euler condition (4.10) for the case of unitary inter-temporal elasticity of substitution. If a steady-state exists, this condition gives the long run rate of return to capital $r^{ss} = \rho + x = r^{k,ss} - \delta$. Then, Equations (5.9) and (5.10) can be used to find the steady-state values p_s^{ss} and $\hat{w}^{ss}$.

To solve for $\hat{k}^{ss}$, proceed as in the preceding chapter noting that the home-good market clearing condition now includes the demand for this sector's output employed as intermediate inputs in other sectors of the economy. For the case of homothetic preferences, $\hat{\epsilon}$ must satisfy

$$\tilde{\epsilon}\left(p_s, \hat{k}\right) \equiv$$

$$\frac{p_s}{\lambda_s}\left(\tilde{y}^s(p_s,\hat{k})\left(1-\sigma_{ss}\right)-\sigma_{sm}\tilde{y}^m(p_s,\hat{k})-\sigma_{sa}\tilde{y}^a\left(p_s\right)H\right) \quad (5.15)$$

Substituting (5.9), (5.10), (5.14) and (5.15) into the budget constraint (4.8) yields the reduced form flow budget constraint as a function of p_s and $\hat{k}$

$$\dot{\hat{k}} = \widetilde{K}(p_s, \hat{k}) \equiv$$

$$\tilde{w}(p_s) + \hat{k}\left[\tilde{r}(p_s) - \delta - x - n\right] + \tilde{\pi}(p_s)H - \tilde{\epsilon}(p_s, \hat{k}) \quad (5.16)$$

To obtain the steady-state value of $\hat{k}$, substitute p_s^{ss} into (5.16), and find the root $\hat{k}^{ss}$ satisfying this equation for $\dot{\hat{k}} = 0$. Knowing $\left(r^{ss}, \hat{w}^{ss}, p_s^{ss}, \hat{k}^{ss}\right)$ allows us to calculate the steady-state levels of the remaining endogenous variables using the intra-temporal equations.[2]

To solve for the transition path, reduce the model to two differential equations in $\hat{k}$ and p_s. Given (5.16), it remains to derive the differential equation for p_s. As in Chapter 4, use the home-good market clearing condition (5.15) and the Euler condition. Differentiate (5.15) with respect to time to obtain

$$\dot{\hat{\epsilon}} = \tilde{\epsilon}_{p_s}\left(p_s, \hat{k}\right)\dot{p}_s + \tilde{\epsilon}_{\hat{k}}\left(p_s, \hat{k}\right)\dot{\hat{k}} \quad (5.17)$$

The Euler condition can be expressed as

$$\dot{\hat{\epsilon}} = \tilde{\epsilon}(p_s, \hat{k})\left(\tilde{r}\left(p_s\right) - \delta - \rho - x\right)$$

Substitute this result into (5.17) to obtain

$$\tilde{\epsilon}\left(p_s, \hat{k}\right)\left(\tilde{r}\left(p_s\right) - \delta - \rho - x\right) = \tilde{\epsilon}_{p_s}\left(p_s, \hat{k}\right)\dot{p}_s + \tilde{\epsilon}_{\hat{k}}\left(p_s, \hat{k}\right)\dot{\hat{k}}$$

[2] The procedure to follow for the case of a non-unitary inter-temporal elasticity and Stone-Geary preferences is more involved but follows the same steps outlined in the previous chapter with the modifications discussed here and is left as an exercise.

Next, substitute (5.16) into the above expression and solve for $\dot{p}_s$ to obtain the second differential equation

$$\dot{p}_s = \frac{[\tilde{r}(p_s) - \delta - \rho - x]\,\tilde{\epsilon}\left(p_s, \hat{k}\right) - \tilde{\epsilon}_{\hat{k}}\left(p_s, \hat{k}\right)\widetilde{K}(p_s, \hat{k})}{\tilde{\epsilon}_{p_s}\left(p_s, \hat{k}\right)} \tag{5.18}$$

Except for the more complicated terms involving $\tilde{\epsilon}\left(p_s, \hat{k}\right)$, this equation is similar to (4.35).

Given the sustainability restriction (4.1) on the rate of productivity growth of H, the two differential equations are autonomous. As such, the time elimination method can be used to solve these two equations numerically to obtain the sequence $\left\{\hat{k}(t), p_s(t)\right\}_{t\in[0,\infty)}$. Substituting this sequence into the reduced form system (5.9), (5.10), (5.11), (5.12), and (5.13) leads to values for all remaining endogenous variables. If Stone-Geary preferences are assumed and $x > 0$, then the system is nonautonomous. Numerical methods for this case are discussed in Chapter 9.

5.1.3 Comparative statics

The comparative static properties of the model are similar to those discussed in the simpler model of Chapter 4. The factor rental rate equations (5.9) and (5.10) are homogeneous of degree one in value-added prices p_{vm} and p_{vs}, and exhibit Stopler-Samuelson like properties. Consider the case where the home-good sector is labor intensive relative to sector m. If capital deepening occurs so that $\dot{r}/r \leq 0$ and $\dot{\hat{w}}/\hat{w} \geq 0$, then the price of the home-good tends to converge from below to its long-run value, that is $\dot{p}_s/p_s \geq 0$. Since the home-good's input-output coefficients, σ_{sj}, are fractions, it follows that the rate of change in value-added prices are $\dot{p}_{vm}/p_{vm} \leq 0$, $\dot{p}_{va}/p_{va} \leq 0$, and $\dot{p}_{vs}/p_{vs} \geq 0$.

Therefore, if home-good production is more labor intensive than manufacturing, in transition growth, the traded good sectors experience a negative internal terms of trade effect as the cost of the intermediate employment of the home-good rises. The relative terms of trade effect between manufacturing and agriculture depends on the share in total cost of the home-good in manufacturing relative to agriculture as suggested by the input-output coefficients σ_{sm} and σ_{sa}. For many countries in the Center for Global Trade Analysis data base,[3] when aggregating national account to the three-sector level (i.e., agriculture, manufacturing, and home-goods), home-good demand is typically the largest off-diagonal element of the input-output matrix of intermediate factor demands. Given the importance of home-goods in traded good production, the negative terms of trade effects discussed in the prior paragraph can be lowered by an increase in home-good productivity. This effect can be interpreted as the releasing of economy-wide resources to the trade competing sectors of the economy. We return to this point later.

As in Chapter 4, the supply functions (5.11) and (5.12) exhibit Rybczynski-like effects. If manufacturing is the most capital intensive sector, holding $\dot{p}_s/p_s = 0$, capital deepening alone causes $\dot{\hat{y}}_m/\hat{y}_m > \dot{\hat{y}}_s/\hat{y}_s$. This capital deepening effect will tend to dominate the negative growth effect on $\hat{y}_m$ from growth in effective labor at rate n since $\dot{k}/k \geq x$. For, $\dot{p}_s/p_s \geq 0$, the home-good market clearing at a higher price allows it to compete for some of the labor and capital that would otherwise be pulled into manufacturing.

The comparative statics of agriculture is indeterminate. Whether the negative effect of $\dot{\hat{w}}/\hat{w} \geq 0$ on profit dominates the positive effect of $\dot{r}/r \leq 0$ in (5.14), depends upon the share of labor in total agricultural costs relative to capital in total costs, and on the rate of change in factor prices. Moreover, the effects of these changes on agricultural supply and factor demand over time can be non-monotonic. In long-run equilibrium, all non-normalized sector outputs grow at the exogenous rate $x + n$.

---

[3]Consult the Center for Global Trade Analysis web site at www.gtap.org. The Center maintains a data base that can be used to construct input-output tables for 113 regions and up to 57 commodities.

5.2 Vertical market structures

The various stages through which final good production passes is of interest for reasons other than the possibility that a more detailed modeling of structure may increase the explanatory power of the model. A stage of production may inhibit the growth in output of one or more stages of production along a production and marketing chain. If an intermediate stage of production faces increased production cost, then its derived demand for inputs from previous stages of production can lead to lower factor payments up-stream – while constrained output supply can lead to higher output prices down-stream.

For example, consider coffee production. Farmers produce coffee cherries, which are sold to firms who process the cherries into coffee beans for export. Since coffee cherries are perishable, international trade in cherries is virtually non-existent, and hence, cherry prices are determined in the domestic market. It follows that the cost of processing cherries into beans affects the price received by farmers. With capital deepening, if the processing stage is labor intensive, cherry processing costs will increase over time as wages rise placing a downward pressure on the price of cherries received by farmers. Whether the production of cherries rises or falls depends upon the share of labor and capital in the total cost of cherry production.

Modeling vertically linked markets is a special case of the intermediate factor structure discussed above. To better illustrate the key features of a vertical market structure, the modifications to the basic model of Chapter 4 are kept to a minimum. As in the previous section, the modifications do not affect the structure of the household problem. We therefore begin by specifying the firm problem.

5.2.1 Firms

The environment for manufacturing and the home-good sector remain unchanged from the basic model. The agricultural sector is divided into two subsectors: primary agriculture, and food processing and retailing. Primary agricultural output is an intermediate input demanded by the food processing and retailing subsector. We assume farmers do not have access to world markets. In such a case, the farm level price p_a becomes an endogenous variable. Food and manufacturing are traded goods, with the exogenous world price of food and manufacturing denoted p_b and p_m, respectively.

Firms producing the manufactured and the home-good are assumed to employ technologies satisfying Assumption 2.2 in the primary factors, labor and capital. Their cost of production $C^j(\hat{w}, k)\,\hat{y}_j$ per effective worker is unchanged from the basic model presented in Chapter 4. The technology of primary agriculture with a sector specific factor results in the same form of the profit function per effective worker as the basic model, $\pi^a(p_a, \hat{w}, r)\,H$.

In this case, the farm level price p_a is determined by the market between farmers and the food processing-retailing sector. The food processing-retailing sector produces food Y_b using labor L_b, capital K_b and primary agricultural output Y_a, and a constant returns to scale technology

$$Y_b = \mathcal{F}^b\left(\mathcal{A}(t)\,L_b, K_b, Y_a\right) \tag{5.19}$$

satisfying Assumption 2.2. The corresponding cost function per effective worker is

$$C^b\left(\hat{w}, r^k, p_a\right)\hat{y}_b \equiv$$

$$\min_{l_b, \hat{k}_b, \hat{y}_a}\left\{\hat{w}l_b + r^k\hat{k}_b + p_a\hat{y}_a : \hat{y}_b \leq f^b\left(l_b, \hat{k}_j, \hat{y}_a\right)\right\} \tag{5.20}$$

where, $\hat{y}_a = Y_a/\mathcal{A}(t)\,L$, $\hat{y}_b = Y_b/\mathcal{A}(t)\,L$, and $l_b = L_b/L$.

5.2.2 *Intra-temporal equilibrium*

An intra-temporal equilibrium for this problem is characterized by the following seven-tuple sequence of positive values

$$\left\{\hat{w}(t), r^k(t), \hat{y}_m(t), \hat{y}_s(t), \hat{y}_b(t), p_{a(t)}, p_s(t)\right\}_{t\in[0,\infty)}$$

satisfying the following seven equations for each t, and capital stock and expenditure pair in the sequence $\left\{\hat{k}(t), \hat{\epsilon}(t)\right\}_{t\in[0,\infty)}$

- zero profits in sectors m, s and b

$$C^j\left(\hat{w}, r^k\right) = p_j,\ j = m, s \tag{5.21}$$

$$C^b\left(\hat{w}, r^k, p_a\right) = p_b \tag{5.22}$$

- labor market clearing

$$\sum_{j=m,s} C^j_{\hat{w}}\left(\hat{w}, r^k\right)\hat{y}_j + C^b_{\hat{w}}\left(\hat{w}, r^k, p_a\right)\hat{y}_b - \boldsymbol{\pi}^a_{\hat{w}}\left(p_a, \hat{w}, r^k\right) H = 1 \tag{5.23}$$

- capital market clearing

$$\sum_{j=m,s} C^j_{r^k}\left(\hat{w}, r^k\right)\hat{y}_j + C^b_{r^k}\left(\hat{w}, r^k, p_a\right)\hat{y}_b - \boldsymbol{\pi}^a_{r^k}\left(p_a, \hat{w}, r^k\right) H = \hat{k} \tag{5.24}$$

- clearing of the intermediate farm to food processing-retailing market

$$C^b_{p_a}\left(\hat{w}, r^k, p_a\right)\hat{y}_b = \boldsymbol{\pi}^a_{p_a}\left(p_a, \hat{w}, r^k\right) H \tag{5.25}$$

- and home-good market clearing

$$\frac{\partial \mathcal{E}\left(p_b, p_s\right)\hat{q}}{\partial p_s} = \hat{y}_s \tag{5.26}$$

The expenditure function

$$\hat{\epsilon} = \mathcal{E}\left(p_b, p_s\right)\hat{q} \tag{5.27}$$

is now expressed in terms of the price of food p_b which replaces p_a in our previous model.

Equations (5.21), (5.22), (5.23), (5.24), (5.25), and (5.26) differ from the Chapter 4 model characterization of equilibrium in three ways. First, the second stage of production (5.19) introduces an additional zero profit condition (5.22) and adds an additional source of factor demand to (5.23) and (5.24). Next, it adds the intermediate good market clearing condition (5.25).

Finally, it adds an additional endogenous price, p_a. The farm level price p_a is endogenous because we are restricting farmers from having access to the world market, while retail establishments face a world price p_b. This system can, in principle, be solved to express the endogenous variables as a function of the exogenous variables (p_m, p_b, H) and the remaining endogenous variables $\left(\hat{k}, \hat{\epsilon}\right)$.

We proceed by using the zero profit conditions (5.21) to express r^k and $\hat{w}$ as functions of p_m (the numeraire price), and p_s to obtain

$$\tilde{w}(p_s) \equiv W(p_m, p_s) \tag{5.28}$$

$$\tilde{r}(p_s) \equiv R(p_m, p_s) \tag{5.29}$$

where $\tilde{w}(p_s)$ and $\tilde{r}(p_s)$ are the zero profit levels of $\hat{w}$ and r^k for prices p_m and p_s.

Substitute these equations for $\hat{w}$ and r^k in (5.22), solve for p_a, and express the result as

$$\tilde{p}^a(p_s) \equiv P^a(p_b, p_m, p_s) \tag{5.30}$$

where it is easily shown that $P^a(\cdot)$ is homogenous of degree one in prices (p_b, p_m, p_s). For the technologies assumed here,[4] $\partial P^a(\cdot)/\partial p_b > 0$, while at least one of the remaining terms is inversely related to changes in p_a.

Substitute (5.28), (5.29), and (5.30) for $\hat{w}$, r^k and p_a in the factor market clearing equations (5.23) and (5.24), and into the intermediate market clearing equation (5.25). The resulting system of equations is linear in $\hat{y}_j$, for $j = b, m, s$. Solving this system leads to the supply functions

$$\hat{y}_j = \tilde{y}^j\left(p_s, \hat{k}\right) \equiv y^j\left(p_b, p_m, p_s, \hat{k}, H\right), \; j = b, m, s \tag{5.31}$$

These functions have the usual property of homogeneity of degree zero in prices. Farm level supply can be expressed as

$$\hat{y}_a \equiv \tilde{y}^a(p_s) H \equiv \boldsymbol{\pi}^a_{p_a}\left(\tilde{p}^a(p_s), \tilde{w}(p_s), \tilde{r}(p_s)\right) H \tag{5.32}$$

[4]For the case of Cobb-Douglas technologies, the elasticity of p_a to the percentage change in p_b is equal to the reciprocal of the cost share of the agricultural input in the total cost of food processing.

For later reference, denote the value-added by H as

$$\tilde{\pi}(p_s)\,H \equiv \boldsymbol{\pi}^a\left(\tilde{p}^a(p_s), \tilde{w}(p_s), \tilde{r}(p_s)\right)H \tag{5.33}$$

5.2.3 Inter-temporal equilibrium

The Euler condition (4.10) remains unchanged. If a steady-state exists, use the long run rate of return to capital condition, $r^{k,ss} = \rho + x + \delta$, and (5.29) to solve for p_s^{ss}. Then, obtain $\hat{w}^{ss}$ and p_a^{ss} using (5.28) and (5.30), respectively.

For the case of homothetic preferences, we once again draw upon the home-good market clearing equation (5.26) to express expenditure $\hat{\epsilon}$ per effective worker as

$$\tilde{\epsilon}\left(p_s, \hat{k}\right) = \frac{p_s}{\lambda_s}\tilde{y}^s\left(p_s, \hat{k}\right) \tag{5.34}$$

Substituting (5.28), (5.29), (5.33) and (5.34) into the budget constraint, we obtain its reduced form

$$\dot{\hat{k}} = \tilde{K}(p_s, \hat{k}) \equiv \tilde{w}(p_s) + \hat{k}\left[\tilde{r}(p_s) - \delta - x - n\right] + \tilde{\pi}(p_s)\,H - \tilde{\epsilon}\left(p_s, \hat{k}\right) \tag{5.35}$$

To solve for the steady-state value of $\hat{k}$, substitute p_s^{ss} into (5.35), and find the root $\hat{k}^{ss}$ satisfying this equation for $\dot{\hat{k}} = 0$. Knowing $\left(r^{k,ss}, \hat{w}^{ss}, p_a^{ss}, p_s^{ss}, \hat{k}^{ss}\right)$ permits the calculation of the remaining endogenous variables.

The transition path is readily reduced to two differential equations in the endogenous variables $\hat{k}$ and p_s. To obtain the differential equation for p_s, time differentiate (5.34) and refer to the result as

$$\dot{\hat{\epsilon}} = \tilde{\epsilon}_{p_s}\left(p_s, \hat{k}\right)\dot{p}_s + \tilde{\epsilon}_{\hat{k}}\left(p_s, \hat{k}\right)\dot{\hat{k}} \tag{5.36}$$

As in the prior section, substitute expression (5.34) into the Euler condition. Then, substitute that result for $\dot{\hat{\epsilon}}$ in (5.36).

Rearranging terms gives

$$\dot{p}_s = \frac{\left[\tilde{r}\left(p_s\right) - \delta - \rho - x\right] \tilde{\epsilon}\left(p_s, \hat{k}\right) - \tilde{\epsilon}_{\hat{k}}\left(p_s, \hat{k}\right) \tilde{K}(p_s, \hat{k})}{\tilde{\epsilon}_{p_s}\left(p_s, \hat{k}\right)}$$

If a steady-state exists, the numerator is zero. The above expression is similar to Equation (5.18).

5.2.4 *An alternative specification*

Many retail services require the physical presence of a buyer and a seller and consequently markets at this level may be treated as home-good markets. Within this context, assume now that farmers face a given world price p_a while the retail price p_b is determined endogenously in the domestic market. In this case the farm level market clearing condition (5.25) is replaced by a market clearing equation at the consumer-retail level,

$$\frac{\partial \mathcal{E}\left(p_b, p_s\right) \hat{q}}{\partial p_b} = \frac{\lambda_b \hat{\epsilon}}{p_b} = \hat{y}_b \tag{5.37}$$

Since the assumptions on technologies and competition remain unchanged, expressions (5.21) and (5.22) still characterize the zero profit conditions for manufacturing and home-good production. The only change is that p_b in expression (5.22) is now endogenous and p_a is exogenous. Substitute (5.28) and (5.29) into (5.22) for $\hat{w}$ and r^k to obtain

$$p_b = \tilde{p}^b\left(p_s\right) \equiv P^b\left(p_m, p_{a,} p_s\right) \tag{5.38}$$

which is the reduced form for the retail price of food p_b. As in the case of (5.30), $P^b\left(\cdot\right)$ is homogeneous of degree one in prices, and for the technology assumed here, $\partial P^b\left(\cdot\right)/\partial p_a > 0$, while at least one of the remaining price effects on p_b are negative.

The factor market clearing conditions (5.23) and (5.24) remain unchanged. Substitute (5.28) and (5.29) for $\hat{w}$, r^k into these two equations and denote the result for the labor and capital market as,

$$\begin{aligned}
\sum_{j=m,s} \tilde{C}^{j}_{\hat{w}}\left(p_s\right)\hat{y}_j + \tilde{C}^{b}_{\hat{w}}\left(p_s\right)\hat{y}_b - \tilde{\boldsymbol{\pi}}^{a}_{\hat{w}}\left(p_s\right)H &= 1 \\
\sum_{j=m,s} \tilde{C}^{j}_{r^k}\left(p_s\right)\hat{y}_j + \tilde{C}^{b}_{r^k}\left(p_s\right)\hat{y}_b - \tilde{\boldsymbol{\pi}}^{a}_{r^k}\left(p_s\right)H &= \hat{k}
\end{aligned}$$

respectively. Again,

$$\tilde{C}^{j}_{i}\left(p_s\right) = C^{j}_{i}\left(W\left(p_m, p_s\right), R\left(p_m, p_s\right)\right), \quad i = \hat{w}, r^k$$

and $\tilde{\pi}^{a}_{i}\left(p_s\right)H$ has an analogous definition.

To solve for the three supply variables $\hat{y}_m$, $\hat{y}_b$ and $\hat{y}_s$, use the two retail market clearing conditions (5.26) and (5.37). Assuming homothetic preferences, these equations allow expressing $\hat{y}_b$ as

$$\hat{y}_b = \frac{\lambda_b}{p_b}\left(\frac{p_s}{\lambda_s}\hat{y}_s\right)$$

These equations are linear in $\hat{y}_j$, and readily expressed as a function of the remaining variables in this three equation system. To solve for the steady-state level of each endogenous variable, and to derive the inter-temporal equilibrium, follow the same reasoning outlined in the prior section. This task is left as an exercise.

5.3 Composite capital

In Chapters 3 and 4, manufacturing output, at a numeraire price of unity, is a final good and an investment good that contributes to the country's stock of capital in which case it depreciates at rate δ each period. The agricultural good and home-good are pure consumption goods and can be viewed as depreciating completely in a single period.

The composite capital concept provides a relatively simple way to allow agricultural and home-goods to also serve as both a consumption and an investment good. The challenge is to incorporate these outputs in a variable that functions analogously to the capital stock in the prior models without introducing an additional state variable, and to do so in a way that accommodates the basic principles of optimization. We accomplish this by requiring the various sectoral outputs to be combined into a composite capital stock variable in a least cost manner.

The capital stock notion presented below achieves these goals, but does so at a cost; the price of the newly defined capital variable is no longer constant, but evolves over time. Hence, the value of capital as an asset is now the product of the price of capital and the level of capital stock .

5.3.1 Asset pricing and the Euler condition

Many authors have discussed asset pricing and the user cost of capital. As such, the discussion below is rather terse. We refer the reader to Romer(2005), Barro and Sala-i-Martin (2004), and Sargent (1979) for a more in depth treatment of this subject.

Consider the asset defined as

$$A(t) = p_k(t) K(t)$$

where p_k is the unit price of capital and K is the capital stock level. If r is the risk free rate of return, what is the opportunity cost of the asset if sold? If the agent sells one unit of the asset, then the risk free reward is rp_k.

Suppose the agent retains one unit of the asset – what is its return? There are three determining factors: (i) the marginal product of capital, which in competitive markets equals r^k less the rate of depreciation δ; (ii) the depreciation of K with unit cost δp_k; and (iii) the change in asset value, $\dot{p}_k$. In equilibrium, an agent is indifferent between selling and retaining the asset.

Putting these considerations together, the value of selling one unit of the asset is equal to the value of retaining the asset, that is:

$$rp_k = r^k - \delta p_k + \dot{p}_k$$

Thus, with perfectly competitive markets, the equilibrium rate of return to the agent should be

$$r = \frac{r^k}{p_k} - \delta + \frac{\dot{p}_k}{p_k} \tag{5.39}$$

This no-arbitrage condition indicates that the difference between r and r^k is no longer δ as in the prior models.

Romer approaches the problem somewhat differently but arrives at the same result. He lists

rp_k	:	interest forgone if firm sells one unit of K, where r is the return to government bonds
δp_k	:	value forgone due to depreciation
$\dot{p}_k$	:	appreciation or depreciation of the price of capital

which leads to the result that the firm's opportunity cost is

$$r^k = rp_k + \delta p_k - \dot{p}_k$$

Solving for r yields the same result as in (5.39).

Consider next the budget constraint in per worker terms

$$\frac{\dot{p}_k K + p_k \dot{K}}{L} = w + rp_k \frac{K}{L} + \frac{\boldsymbol{\pi}^a (p_a, w, r) \mathcal{B}(t)}{L} H - \frac{E}{L}$$

Growth in the number of workers at rate n implies

$$\dot{p}_k k + p_k \dot{k} + p_k k n = w + rp_k k + \boldsymbol{\pi}^a (p_a, w, r) \tilde{\mathcal{B}}(t) H - \epsilon \tag{5.40}$$

Here, $\tilde{\mathcal{B}}(t) = \mathcal{B}(t) / L(t)$, $\epsilon = \mathcal{E}(p_a, p_s) q$ is expenditure per worker, and p_m remains normalized to unity.

Substitute (5.39) into (5.40) and simplify to get

$$\dot{k} = \frac{1}{p_k} \left[w + r^k k + \boldsymbol{\pi}^a (p_a, w, r) \tilde{\mathcal{B}}(t) H - \epsilon \right] - k(\delta + n) \tag{5.41}$$

Then, the budget constraint expressed in effective units of labor is

$$\dot{\hat{k}} = \frac{1}{p_k} \left[\hat{w} + r^k \hat{k} + \boldsymbol{\pi}^a (p_a, \hat{w}, r) H - \hat{\epsilon} \right] - \hat{k}(x + \delta + n) \tag{5.42}$$

Form the present value Hamiltonian, and derive the Euler condition in the same manner as in Chapter 4. Notice, however, the presence of p_k which evolves with time. The reader is left to verify that the resulting Euler condition for $1/\theta < 1$ is

$$\frac{\dot{\epsilon}}{\epsilon} = \frac{1}{\theta}\left[\frac{r^k}{p_k} - \delta - \rho - (1-\theta)\lambda\frac{\dot{p}_s}{p_s} + \frac{\dot{p}_k}{p_k}\right]$$

and for unitary elasticity of substitution we have

$$\frac{\dot{\epsilon}}{\epsilon} = \frac{r^k}{p_k} - \delta - \rho + \frac{\dot{p}_k}{p_k} \tag{5.43}$$

In terms of effective labor units, (5.43) becomes,

$$\frac{\dot{\hat{\epsilon}}}{\hat{\epsilon}} = \frac{r^k}{p_k} - \delta - \rho - x + \frac{\dot{p}_k}{p_k} \tag{5.44}$$

Households now take into account the price of the asset p_k and its evolution over time in choosing their path of expenditures.

5.3.2 *Specification of composite capital*

We model composite capital by combining the incremental outputs of all three-sectors of the economy in a least cost manner, presuming some underlying technology. For purposes here, we choose a constant returns to scale Cobb-Douglas technology

$$F\left(y_{mk}(t), y_{ak}(t), y_{sk}(t)\right)$$

with production elasticities $\lambda_{jk} > 0$, and $\Sigma_j \lambda_{jk} = 1$, $j = a, m, s$.[5] Each y_{jk} is the quantity of sector j output per worker allocated to the production of an increment of composite capital per worker. Then, suppressing t, the minimum cost of producing a unit of composite capital is

$$c^k\left(p_m, p_a, p_s\right) \equiv \min_{y_{mk}, y_{ak}, y_{sk}} \left\{\sum_{j=1}^{3} p_j y_{jk} : 1 \leq F\left(y_{mk}, y_{ak}, y_{sk}\right)\right\} \tag{5.45}$$

[5] The Leontief and the constant elasticity of substitution function are alternative specifications.

where $c^k(\cdot)$ is non-decreasing and homogeneous of degree one in all prices p_j. In equilibrium, zero profit in composite capital production requires $p_k(t) = c^k(p_m, p_a, p_s(t))$, and hence, defines the equilibrium unit price of composite capital at each point in time.

Since the prices p_m and p_a are exogenous, the evolution of p_k is determined by p_s. Shephard's lemma applied to $c^k(\cdot)$ yields the cost minimizing amount of y_{jk} used to produce a unit of capital. At each t, the minimum total cost per worker of producing "new" capital is

$$c^k(p_m, p_a, p_s)\left[\dot{k} + k(\delta + n)\right]$$

Here, $p_k k\delta$ is the cost of replacing depreciated capital, $p_k kn$ is the cost of adding capital to accommodate new workers, and $p_k\dot{k}$ is the cost of increasing the capital stock per worker.

Note, the rate of change in the price of composite capital is equal to the product of two terms: (i) λ_{sk}, the share of the home-good in the total cost of composite capital, and (ii) $\dot{p}_s/p_s$, the rate of change in the price of the home-good. That is,

$$\frac{\dot{p}_k}{p_k} = \frac{d}{dt}(\log p_k) = \frac{c^k_{p_s}(p_m, p_a, p_s)\, p_s}{p_k}\frac{\dot{p}_s}{p_s} = \lambda_{sk}\frac{\dot{p}_s}{p_s} \tag{5.46}$$

Below, we substitute $\lambda_{sk}(\dot{p}_s/p_s)$ for $\dot{p}_k/p_k$ in the Euler condition (5.44).

In units of effective labor, the total amount of the home-good per effective worker allocated to composite capital is given by

$$\begin{aligned}\hat{y}_{sk} &= \tilde{y}^{sk}(p_s)\left[\dot{\hat{k}} + \hat{k}(x + n + \delta)\right] \\ &= \lambda_{sk}\frac{p_k}{p_s}\left[\dot{\hat{k}} + \hat{k}(x + n + \delta)\right]\end{aligned} \tag{5.47}$$

The remaining components of capital are

$$\hat{y}_{jk} = \tilde{y}^{jk}(p_s)\left[\dot{\hat{k}} + \hat{k}(x + n + \delta)\right]$$

$$= \lambda_{jk} \frac{p_k}{p_j} \left[\dot{\hat{k}} + \hat{k} \left(x + n + \delta \right) \right], \quad j = a, m \tag{5.48}$$

We next discuss the model's equilibrium conditions.

5.3.3 *Intra-temporal equilibrium*

The characterization of intra-temporal equilibrium is mostly unchanged from the basic model in Chapter 4. The zero profit conditions for production of the manufacturing and the home-good yield the familiar rental rate equations

$$\tilde{w}\left(p_s\right) \equiv W\left(p_m, p_s\right) \tag{5.49}$$

$$\tilde{r}\left(p_s\right) \equiv R\left(p_m, p_s\right) \tag{5.50}$$

As in Chapter 4, substitute (5.49) and (5.50) into the factor market clearing equations and solve for $\hat{y}_m$ and $\hat{y}_s$ to get the reduced form supply functions

$$\begin{aligned} \hat{y}_j &= \tilde{y}^j\left(p_s, \hat{k}\right) \equiv y^j\left(p_m, p_a, p_s, \hat{k}, H\right), \quad j = m, s \\ \hat{y}_a &= \tilde{y}^a\left(p_s\right) H \equiv \boldsymbol{\pi}^a_{p_a}\left(p_a, W\left(p_m, p_s\right), R\left(p_m, p_s\right)\right) H \end{aligned}$$

For later reference, represent that value-added by H as

$$\tilde{\pi}\left(p_s\right) H = \boldsymbol{\pi}^a\left(p_a, \tilde{w}\left(p_s\right), \tilde{r}\left(p_s\right)\right) H \tag{5.51}$$

These reduced form functions have the same properties of the analogous functions in Chapter 4.

The home good market clearing condition, however, is altered because the total demand is now composed of two parts: household consumption demand and the derived demand from composite capital production. Thus the home-good market clearing condition becomes

$$\frac{\lambda_s \hat{\epsilon}}{p_s} = \tilde{y}^s\left(p_s, \hat{k}\right) - \tilde{y}^{sk}\left(p_s\right)\left[\dot{\hat{k}} + \hat{k}\left(x + n + \delta\right)\right]$$

which allows expressing expenditure per effective worker as

$$\hat{\epsilon} = \tilde{\epsilon}\left(p_s, \hat{k}, \dot{\hat{k}}\right) = \frac{p_s}{\lambda_s}\left\{\tilde{y}^s\left(p_s, \hat{k}\right) - \tilde{y}^{sj}\left(p_s\right)\left[\dot{\hat{k}} + \hat{k}\left(x + n + \delta\right)\right]\right\} \tag{5.52}$$

5.3.4 *Inter-temporal equilibrium*

We proceed with the same strategy of using the Euler condition to determine one of the steady-state variables. Substitute (5.45) for p_k, and (5.46) for $\dot{p}_k/p_k$ in the Euler condition (5.44). Expressed in effective labor units, we have

$$\frac{\dot{\hat{\epsilon}}}{\hat{\epsilon}} = \frac{r^k}{\tilde{c}^k\left(p_s\right)} - x - \delta - \rho + \lambda_{sk}\frac{\dot{p}_s}{p_s}, \tag{5.53}$$

where $\tilde{c}^k\left(p_s\right) \equiv \tilde{c}^k\left(p_m, p_a, p_s\right)$. If a steady-state exists, then substituting (5.50) for r^k gives

$$\tilde{r}\left(p_s\right) = \tilde{c}^k\left(p_s\right)\left(x + \delta + \rho\right)$$

The root p_s^{ss} satisfying this equation is the steady-state price of the home-good. This result permits, as in the previous case, the derivation of $r^{k,ss}$, and $\hat{w}^{ss}$ using (5.49) and (5.50).

Next, substitute the steady-state values of $\hat{w}$, r^k and p_s into the budget constraint, (5.42). If a steady-state exists, then $\dot{\hat{k}} = 0$, and the steady-state level of capital per unit of effective labor, $\hat{k}^{ss}$, is the root satisfying

$$\frac{1}{\tilde{c}^k\left(p_s\right)}\left[\hat{w}^{ss} + r^{k,ss}\hat{k} + \tilde{\pi}(p_s^{ss})\,H - \tilde{\epsilon}\left(p_s^{ss}, \hat{k}, 0\right)\right] - \hat{k}(x + \delta + n) = 0$$

As in the previous model, the remaining endogenous variables can be calculated using the reduced form equations derived from the intra-temporal conditions.

We now reduce the model to two differential equations in $\hat{k}$ and p_s. We present this case first, and then briefly discuss an analogous three-differential-equation system. The three-equation system provides the same result as the two-equation system, but the former system requires less substitution of other equations and is useful for representing expenditure systems that are more complex than the system discussed here.

The differential equation for capital is the budget constraint (5.42) in which we replace $\hat{w}$, r^k, and π by (5.49), (5.50), and

(5.51), and $\hat{\epsilon}$ by (5.52) . These substitutions leave $\dot{\hat{k}}$ appearing on both sides of the equation

$$\dot{\hat{k}} = \frac{1}{\tilde{c}^k(p_s)}\left[\tilde{w}(p_s) + \tilde{r}(p_s)\hat{k} + \tilde{\pi}(p_s)H - \tilde{\epsilon}\left(p_s, \hat{k}, \dot{\hat{k}}\right)\right] - \hat{k}(x+\delta+n)$$

Inspection of (5.52) indicates this equation is readily solved for $\dot{\hat{k}}$. Denote the solution as

$$\dot{\hat{k}} = \tilde{K}\left(p_s, \hat{k}\right) \tag{5.54}$$

We follow the same procedures as in the basic model of Chapter 4. Use the market clearing equation for the home-good (5.52) and the Euler condition (5.53) to derive the differential equation for p_s. First, replace $\dot{\hat{k}}$ in (5.52) by (5.54), and express the result as

$$\hat{\epsilon} = \tilde{\epsilon}\left(p_s, \hat{k}, \tilde{K}\left(p_s, \hat{k}\right)\right) = \bar{\epsilon}\left(p_s, \hat{k}\right) \tag{5.55}$$

Differentiate this equation with respect to time,

$$\dot{\hat{\epsilon}} = \frac{d}{dt}\bar{\epsilon}\left(p_s, \hat{k}\right)$$

and replace $\dot{\hat{\epsilon}}$ using (5.53), to obtain

$$\hat{\epsilon}\left[\frac{\tilde{r}(p_s)}{\tilde{c}^k(p_s)} - \delta - \rho - x + \lambda_{sk}\frac{\dot{p}_s}{p_s}\right] = \bar{\epsilon}_{p_s}\left(p_s, \hat{k}\right)\dot{p}_s + \bar{\epsilon}_{\hat{k}}\left(p_s, \hat{k}\right)\dot{\hat{k}}$$

Substitute (5.54) for $\dot{\hat{k}}$ and (5.55) for $\hat{\epsilon}$. The resulting equation

$$\bar{\epsilon}\left(p_s, \hat{k}\right)\left[\frac{\tilde{r}(p_s)}{\tilde{c}^k(p_s)} - \delta - \rho - x + \lambda_{sk}\frac{\dot{p}_s}{p_s}\right]$$
$$= \bar{\epsilon}_{p_s}\left(p_s, \hat{k}\right)\dot{p}_s + \bar{\varepsilon}_{\hat{k}}\left(p_s, \hat{k}\right)\tilde{K}\left(p_s, \hat{k}\right)$$

is linear in $\dot{p}_s$. Solve for $\dot{p}_s$ and express the result as

$$\dot{p}_s = \frac{p_s\left\{\bar{\epsilon}\left(p_s,\hat{k}\right)\left[\frac{\tilde{r}(p_s)}{\tilde{c}^k(p_s)}-\delta-\rho-x\right]-\bar{\epsilon}_{\hat{k}}\left(p_s,\hat{k}\right)\tilde{K}\left(p_s,\hat{k}\right)\right\}}{p_s\bar{\epsilon}_{p_s}\left(p_s,\hat{k}\right)-\lambda_{sk}\bar{\epsilon}\left(p_s,\hat{k}\right)} \tag{5.56}$$

If a steady-state exists, the numerator is zero and hence $\dot{p}_s = 0$. However, the equation is more complex than in the case of the other models presented in this chapter. The numerical solution $\left\{\hat{k}(t), p_s(t)\right\}_{t\in[0,\infty)}$ to (5.55) and (5.56) permits the derivation of the remaining endogenous variables using the intra-temporal equilibrium conditions.

A somewhat less complex system is to reduce the system to three differential equations in $\dot{\hat{k}}$, $\dot{\hat{\epsilon}}$ and $\dot{p}_s$. The same system of equations are used, namely the budget constraint

$$\dot{\hat{k}} = \frac{1}{\tilde{c}^k(p_s)}\left[\tilde{w}(p_s)+\tilde{r}(p_s)\hat{k}+\tilde{\pi}(p_s)H-\hat{\epsilon}\right]-\hat{k}(x+\delta+n)$$

for $\dot{\hat{k}}$, the Euler condition (5.53) for $\dot{\hat{\epsilon}}$ and the derivative of the market clearing equation (5.55) with respect to time for $\dot{p}_s$. Then, substitutions are made to replace the "." variables. The result is three differential equations in the three variables $\hat{k}, \hat{\epsilon}$ and $\hat{p}_s$. In this case, a numerical solution using the time elimination method comprises two policy functions which yield numerical values for $\hat{\epsilon}$ and $\hat{p}$ that are consistent with equilibrium for a sequence of values of $\hat{k}$ over the interval $\hat{k}(0)$ to $\hat{k}^{ss}$. This general procedure is presented in Chapter 9.

5.3.5 Discussion

The effect of composite capital on the economy does not reduce to a simple comparative static analysis. The remarks that can be made depend on changes in the relative magnitudes of variables. That is, the effects are sensitive to the data that characterize the structure of the economy. Complications arise because the price of composite capital enters the Euler condition (5.53) and consequently affects the transition path and the level of the steady-state equilibrium of the economy. Hence, this addition to the model has wide-ranging effects.

The direction of the effects depends upon whether the price of the home-good converges from below or above to its steady-state value, i.e., whether $\dot{p}_s/p_s$ is positive or negative since $\dot{p}_k/p_k = \lambda_{sk}\left(\dot{p}_s/p_s\right)$. Thus the weight, λ_{sk}, of home-good in composite capital matters. The larger is λ_{sk} the greater is the demand per worker for y_{sk} which affects the rate of change in $\dot{p}_s/p_s$ and its steady-state level. If the home-good sector is the most labor intensive, then a rise in p_s and hence p_k increases the home-good sector's demand for resources, especially labor.

The "pulling" of labor into the home-good sector tends to lower the productivity of composite capital in the other sectors and hence their demand for capital. The increased demand for the home-good can cause the steady-state level of the home-good price to exceed that of the case where λ_{sk} is small. As can be seen from the Euler condition, the larger is p_s^{ss} the larger is $r^{k,ss}$. A larger share λ_{sk} thus lowers production in the traded good sectors of the economy. The effect of a larger share λ_{sk} on agricultural output and the employment of labor and capital relative to the industrial sector depends on differences in their factor shares, as well as the share of land rent in total production costs. In the latter case, since $\dot{p}_s/p_s$ must converge to a larger steady-state value than when λ_{sk} is small, agriculture will suffer a relatively larger negative internal terms of trade effect than the industrial sector.

5.4 Government

We return to the basic model of the previous chapter to incorporate government. The World Bank's World Development Indicators data base shows that general government consumption expenditures, as a percent of GDP, typically range from the upper teens to low twenties. These expenditures also tend to be concentrated on the consumption of services. Combining government and household consumption therefore tends to greatly distort actual household expenditure share,[6] while omitting government from a modeled economy leaves a significant percent of the production of services and the allocation of labor to be accounted for in some way. Moreover, taxes, subsidies and fiscal imbalances are typically an important aspect of many policy questions. Other issues also arise, such as fiscal deficits and debt.

The major difficulty in modeling government is that additional notation must be introduced which causes some inevitable awkwardness to model specification. While our approach allows for an algebraic closed form solution of many of the reduced form equations, in empirical application, computer code facilitates these derivations. Nevertheless, the following derivations are essential to guide this process and to understand how taxes and government expenditures affect household and firm behavior.

5.4.1 Government consumption and revenues

To avoid the presumption that government is an optimizing agent, we treat government's propensity to consume as a constant proportion, λ_g, of the returns to primary factors. Thus, we take as given the rule that government expenditures per worker are a constant percent of the value of primary factor payments

$$\epsilon_g = \lambda_g \left(w + r^k k + \pi H \right) \tag{5.57}$$

where π is rent to land per economy-wide worker. The shares of total government expenditure on industrial, agricultural and home-goods, denoted λ_j^g, $j = a, m, s$, are also given by data.

---

[6]More troubling is the implicit assumption that government is solving the same optimization problem as households.

Assuming a constant percentage implies that government consumption of the jth good is given by

$$q_j^g = \frac{\lambda_j^g \epsilon_g}{p_j} \tag{5.58}$$

Revenues flow from indirect production taxes, tariffs, export taxes, and lump-sum transfers from households. We begin with taxes on traded goods.

The domestic price of traded goods are

$$p_j = p_j^w \left(1 + \tau_j\right) \quad j = m, a \tag{5.59}$$

where p_j^w is the world price and τ_j is the ad-valorem tax rate that remains constant over time. If good j is imported, a value of $\tau_j > 0$ is an import duty and $p_j^w < p_j$, and a subsidy if $\tau_j < 0$. If the good is exported, than $\tau_j > 0$ is a subsidy and $p_j^w < p_j$, and a tax otherwise. In the case of traded goods, let $z_j > 0$ denote excess supply per worker and $z_j < 0$ excess demand per worker where

$$z_m \equiv y_m - q_m^g - q_m - \dot{k} - k(n + \delta)$$

and

$$z_a \equiv y_a - q_a^g - q_a$$

Government revenue per worker from foreign trade is the product $p_j^w \tau_j z_j$.

Government can also impose an indirect tax on production. Denote the indirect tax per unit of sector j output by $p_j t_j^I > 0$, or subsidy if t_j^I is negative.

At each t, total government tax revenue per worker is thus

$$G_{rev} = \sum_{j=a,m} p_j^w \tau_j \; z_j + \sum_{j=a,m,s} p_j \tau_j^I y_j \tag{5.60}$$

If government is constrained to balance expenditures with revenues, then lump-sum transfers from households per worker become,

$$T_{gov} = G_{rev} - \epsilon_g \tag{5.61}$$

In this framework, the price of the traded good (5.59) is not necessarily unity, although as we discuss in a later chapter, the model is fit to data so that all domestic base period prices are unity.

In per worker terms, transfers (5.61) appear as a lump-sum term in the household's budget constraint

$$\dot{k} = \frac{1}{p_m}\left(w + r^k k + \pi H + T_{gov} - \epsilon\right) - k\left(n + \delta\right) \tag{5.62}$$

Later, we return to these conditions and link them to the equilibrium properties of the model. Since p_m is constant for all t, the Euler condition for the case of unitary inter-temporal elasticity remains unchanged, although the rate of return r^k is a function of trade and indirect taxes.

5.4.2 Firms

Continuing with the neoclassical technology assumption, the gross value of output equals total costs plus indirect taxes per effective worker

$$p_j \hat{y}_j = \hat{w}\ell_j + r^k \hat{k}_j + p_j \tau_j^I \hat{y}_j, \ j = m, s \tag{5.63}$$

and

$$p_a \hat{y}_a = \hat{w}\left(1 - \ell_m - \ell_s\right) + r^k \hat{k}_a + \hat{\pi} H + p_a \tau_a^I \hat{y}_a \tag{5.64}$$

for the case of agriculture, where ℓ_j denotes the share of the total labor force employed in sector j. The corresponding cost and revenue functions are

$$\left(C^j\left(\hat{w}, r^k\right) + p_j \tau_j^I\right) y_j, \ j = m, s$$

$$\pi e^{-xt} H = \pi^a\left(p_a\left(1 - \tau_a^I\right), \hat{w}, r^k\right) H$$

The trade and indirect taxes affect the incentives of firms through the price received for output. However, the indirect tax does not affect household incentives directly as they perceive market clearing at prices p_j.

5.4.3 Intra-temporal equilibrium

In the first part of this section we proceed to layout the conditions for equilibrium that must hold at each t. We omit a number of intermediate steps since they are similar to the basic model. Then, we derive the reduced form equations that link government revenues and expenditures, (5.60) and (5.61), to these equilibrium conditions.

The characterization of equilibrium

The intra-temporal characterization of equilibrium is virtually identical to the basic model. We restate this as: Given the sequence of capital stock and expenditure pairs, $\left\{\dot{k}(t), \hat{\epsilon}(t)\right\}_{t\in[0,\infty)}$, an intra-temporal equilibrium for each t is the five-tuple sequence of positive values

$$\left\{\hat{w}(t), r^k(t), \hat{y}_m(t), \hat{y}_s(t), p_s(t)\right\}_{t\in[0,\infty)}$$

that satisfy the following five equations

- zero profits in sectors $j = m, s$

$$C^j\left(\hat{w}, r^k\right) = p_j\left(1-\tau_j^I\right),\ j = m, s \tag{5.65}$$

- labor market clearing

$$\sum_{j=m,s} C^j_{\hat{w}}\left(\hat{w}, r^k\right)\hat{y}_j - \boldsymbol{\pi}^a_{\hat{w}}\left(p_a\left(1-\tau_a^I\right), \hat{w}, r^k\right)H = 1 \tag{5.66}$$

- clearing of the capital market

$$\sum_{j=m,s} C^j_{r^k}\left(\hat{w}, r^k\right)\hat{y}_j - \boldsymbol{\pi}^a_{r^k}\left(p_a\left(1-\tau_a^I\right), \hat{w}, r^k\right)H = \hat{k} \tag{5.67}$$

- and clearing of the market for the home-good

$$\frac{\partial\mathcal{E}\left(p_a, p_s\right)\hat{q}}{\partial p_s} = \hat{y}_s - \hat{q}^g_s \tag{5.68}$$

Account is now taken of government consumption of the home-good, $\hat{q}_s^g = q_s^g e^{-xt}$, using (5.58). The difference between this characterization and that of the previous chapter is mostly in the definition of prices (5.59), the role of indirect taxes and government consumption in (5.68).

Proceeding as in the previous model, the zero profit conditions imply $\hat{w}$ and r^k can be expressed by the following two reduced forms

$$\tilde{w}(p_s) \equiv W\left(p_m\left(1-\tau_m^I\right), p_s\left(1-\tau_s^I\right)\right) \quad (5.69)$$
$$\tilde{r}(p_s) \equiv R\left(p_m\left(1-\tau_m^I\right), p_s\left(1-\tau_s^I\right)\right) \quad (5.70)$$

As above, we use the $\tilde{w}(\cdot)$ notation. Plug (5.69) and (5.70) into (5.66) and (5.67) for $\hat{w}$ and r^k and solve the resulting two equations to obtain the reduced form supply functions

$$\hat{y}_j = \tilde{y}^j\left(p_s, \hat{k}\right) \equiv y^j\left(p_m(1-\tau_m^I), p_a(1-\tau_a^I), p_s(1-\tau_s^I), \hat{k}, H\right) \quad (5.71)$$

for $j = m, s$. For agricultural supply, proceed similarly, and express the result as

$$\hat{y}_a = \tilde{y}^a(p_s) = \pi^a_{p_a(1-\tau_a^I)}\left(p_a\left(1-\tau_a^I\right), \tilde{w}(p_s), \tilde{r}(p_s)\right) H \quad (5.72)$$

For later reference, let

$$\tilde{\pi}(p_s) H = \boldsymbol{\pi}^a\left(p_a\left(1-\tau_a^I\right), \tilde{w}(p_s), \tilde{r}(p_s)\right) H \quad (5.73)$$

The supply system (5.71) and (5.72) thus reduces to the familiar basic structure of Chapter 4.

The next step is to derive the function for transferring government revenue to or from households, Equation (5.61), into variables p_s,and $\hat{k}$ that are common to the above reduced form equations.

Government expenditures, revenues and transfers

We begin first with government expenditures. Sum the identities (5.63) and (5.64) to derive aggregate GDP per effective worker. Doing so, and using the factor rental rate equations (5.69) and

(5.70), and the supply functions (5.71) and (5.72), GDP per effective worker is

$$\begin{aligned}\tilde{G}\left(p_s, \hat{k}\right) &\equiv \sum_{j=m,s} p_j \left(1 - it_j\right) \tilde{y}^j \left(p_s, \hat{k}\right) + p_a \left(1 - \tau_a^I\right) \tilde{y}^a \left(p_s\right) \\ &= \tilde{w}\left(p_s\right) + \tilde{r}\left(p_s\right) \hat{k} + \tilde{\pi}\left(p_s\right) H \qquad (5.74)\end{aligned}$$

Hence, government expenditure (5.57), in per effective worker terms becomes

$$\hat{\epsilon}_g = \lambda_g \tilde{G}\left(p_s, \hat{k}\right) \qquad (5.75)$$

which permits the derivation of government consumption using (5.58).

To derive government revenues, it remains to define the excess supply and demand for traded goods in (5.60). Given household preferences are homothetic, the excess supply and demand in manufacturing is equal to supply less: (i) the consumption demand of households and government and (ii) the amount of the industrial good allocated to investment. In effective labor terms we have:

$$\tilde{y}^m \left(p_s, \hat{k}\right) - \frac{\lambda_m \hat{\epsilon}}{p_m} - \frac{\lambda_m^g \lambda_g \tilde{G}\left(p_s, \hat{k}\right)}{p_m} - \dot{\hat{k}} - \hat{k}\left(x + n + \delta\right) \qquad (5.76)$$

and for agriculture

$$\tilde{y}^a \left(p_s\right) - \frac{\lambda_a \hat{\epsilon}}{p_a} - \frac{\lambda_a^g \lambda_g \tilde{G}\left(p_s, \hat{k}\right)}{p_a} \qquad (5.77)$$

Finally, it is useful to express (5.76) and (5.77) in terms p_s, and $\hat{k}$ which requires a reduced form equation for $\hat{\epsilon}$. Note that it follows from home-good market clearing (5.68) that household expenditures can be expressed as

$$\tilde{\epsilon}\left(p_s, \hat{k}\right) = \frac{p_s}{\lambda_s}\left[\tilde{y}^s \left(p_s, \hat{k}\right) - \frac{\lambda_s^g \lambda_g \tilde{G}\left(p_s, \hat{k}\right)}{p_s}\right] \qquad (5.78)$$

Thus, in effective labor units, net trade of the manufactured good is given by

$$\tilde{z}^m\left(p_s,\hat{k},\dot{\hat{k}}\right)\equiv$$

$$\tilde{y}^m\left(p_s,\hat{k}\right)-\frac{\lambda_m\tilde{\epsilon}\left(p_s,\hat{k}\right)}{p_m}-\frac{\lambda_m^g\lambda_g\tilde{G}\left(p_s,\hat{k}\right)}{p_m}-\dot{\hat{k}}-\hat{k}\left(x+n+\delta\right) \tag{5.79}$$

and for agriculture we have

$$\tilde{z}^a\left(p_s,\hat{k}\right)\equiv\tilde{y}^a\left(p_s\right)-\frac{\lambda_a\tilde{\epsilon}\left(p_s,\hat{k}\right)}{p_a}-\frac{\lambda_a^g\lambda_g\tilde{G}\left(p_s,\hat{k}\right)}{p_a} \tag{5.80}$$

Notice the presence of $\dot{\hat{k}}$ in (5.79). This term will be dealt with when the household's budget constraint is specified.

To summarize this section, we have now derived the government's revenue function

$$\tilde{G}^{rev}\left(p_s,\hat{k},\dot{\hat{k}}\right)=p_m^w\tau_m\tilde{z}^m\left(p_s,\hat{k},\dot{\hat{k}}\right)+p_a^w\tau_a\tilde{z}^a\left(p_s,\hat{k}\right)+$$

$$\sum_{j=m,s}p_j\tau_j^I\tilde{y}^j\left(p_s,\hat{k}\right)+p_a\tau_a^I\tilde{y}^a\left(p_s\right) \tag{5.81}$$

The government's expenditure function is (5.75). Consequently, for each t, the transfers to or from households to assure that the government's fiscal accounts – per effective worker – balance is equal to

$$\tilde{T}^{gov}\left(p_s,\hat{k},\dot{\hat{k}}\right)=\lambda_g\tilde{G}\left(p_s,\hat{k}\right)-\tilde{G}^{rev}\left(p_s,\hat{k},\dot{\hat{k}}\right) \tag{5.82}$$

5.4.4 Inter-temporal equilibrium

The household optimization problem remains unchanged from the basic model. While the budget constraint (5.62) contains the transfer term (5.82), the Euler condition is unaffected as the transfers are treated as lump-sum by the household. Consequently, we proceed with the budget constraint, and substitute the reduced form equations in the previous section to obtain

$$\dot{\hat{k}} + \hat{k}(x + n + \delta) =$$

$$\frac{1}{p_m}\left[\tilde{w}(p_s) + \tilde{r}(p_s)\hat{k} + \tilde{\pi}(p_s)H + \tilde{T}^{gov}\left(p_s, \hat{k}, \dot{\hat{k}}\right) - \tilde{\epsilon}\left(p_s, \hat{k}\right)\right]$$

Since, by (5.79), $\tilde{T}^{gov}\left(p_s, \hat{k}, \dot{\hat{k}}\right)$ is linear in $\dot{\hat{k}}$, the budget constraint can be solved for this term. The result is the differential equation which we simply state as

$$\dot{\hat{k}} = \tilde{K}\left(p_s, \hat{k}\right) \tag{5.83}$$

Again, we return to the Euler condition and the home-good market clearing condition (5.78) to derive the equation for $\dot{p}_s$. Differentiate this equation with respect to time, and refer to the result as

$$\dot{\hat{\epsilon}} = \tilde{\epsilon}_{p_s}\left(p_s, \hat{k}\right)\dot{p}_s + \tilde{\epsilon}_{\hat{k}}\left(p_s, \hat{k}\right)\dot{\hat{k}} \tag{5.84}$$

Proceed in the same manner as in the previous derivations; use the home-good market clearing condition (5.78) to substitute for $\hat{\epsilon}$ in the Euler condition

$$\dot{\hat{\epsilon}} = \tilde{\epsilon}\left(p_s, \hat{k}\right)\left(\frac{r^k}{p_m} - \delta - \rho - x\right)$$

Then, use this result to substitute for $\dot{\hat{\epsilon}}$ in (5.84). Finally, use (5.83) to substitute for $\dot{\hat{k}}$, and solve for $\dot{p}_s$ to obtain the differential equation

$$\dot{p}_s = \frac{\left[\tilde{r}(p_s)/p_m - \delta - \rho - x\right]\tilde{\epsilon}\left(p_s, \hat{k}\right) - \tilde{\epsilon}_{\hat{k}}\left(p_s, \hat{k}\right)\tilde{K}(p_s, \hat{k})}{\tilde{\epsilon}_{p_s}\left(p_s, \hat{k}\right)}$$

The steady-state equilibrium is also derived as in the base model. Use the Euler condition and the factor rental rate equation (5.83) to obtain $r^{k,ss}$ and p_s^{ss} from which $\hat{w}^{ss}$ is calculated. Set $\dot{\hat{k}}$ to zero and use the budget constraint along with the previously determined steady-state values to solve for $\hat{k}^{ss}$. The steady-state values of all remaining endogenous variables can now be calculated using the intra-temporal conditions.

The effects of taxes, transfers and government consumption are complex because of the various channels through which they affect behavior. We thus delay a discussion of some of these effects at this point and return to them when we discuss the empirical example in Chapter 6.

5.5 A numerical example

The numerical example adds intermediate inputs to the empirical example of Chapter 4. Chapter 8 provides details of the social accounting matrix upon which the parameters of this example are based. We discuss and contrast the results of this section with the results from the preceding example. We conclude with a simulation in which a positive one-time change to the scale parameter of the service sector production function is shown to cause relatively large multiplier effects on the rest of the economy and to increase foreign trade.

5.5.1 *Parameter estimation*

The parameters reported in Table 4.1 remain unchanged. New are the levels of the agricultural, industrial and service good output employed as intermediate inputs of production. The output of each sector is now expressed in terms of gross output with the value-added by each sector unchanged from the previous example. The shares of gross output accounted for by intermediate inputs are reported in Table 5.1.

Two features are common to the input-output data of many countries. The first is the relatively large share of the value of

Table 5.1 Factor share in gross output

	Intermediate input to		
	Industry	Agriculture	Service
Industry	0.474	0.113	0.149
Agriculture	0.015	0.233	0.015
Services	0.178	0.136	0.197
Total	0.666	0.481	0.361

Source: The GTAP data set

gross output accounted for by intermediate inputs. Intermediate inputs account for over 66 percent of the value of the industrial sector's gross output, followed by agriculture at 48 percent and the service sector at 36 percent. Second, own output comprises the largest share of total intermediate inputs employed. Another relatively common feature is the service sector share of gross output is often the largest off-diagonal element of a matrix of intermediate input shares. The service sector accounts for almost 18 percent of the value of industrial sector gross output and 14 percent of the value of agriculture's gross output.

The basic economic forces underlying the transition to long-run growth discussed in the empirical example of Chapter 4 prevail here. A difference is the negative internal terms of trade effect the increase in the service good price has on traded goods production. Compared to agriculture, the marginally higher share of service good inputs in the industrial sector's gross output suggests a rise in the price of the service good will have a larger negative effect on the cost of producing the industrial good than it will on the cost of producing the agricultural good.

5.5.2 Empirical results

For the year 2002, the model predicts a rate of growth of GDP per worker equal to 3.03 percent; this is 0.05 percentage points less than that found in the example in Chapter 4. By 2013, the rate of growth in GDP per worker is equal to 2.5 percent. This rate of growth represents the economy's half-way point to steady-state growth in GDP per worker of 1.9 percent. Income per worker doubles in about 28 years and triples by 2049. By the year 2050, the economy is within 3.7 percent of its long-run rate of growth in GDP per worker. In the empirical model of Chapter 4, by 2050 the economy was within 5.2 percent of its long-run rate of growth in per worker GDP.

Table 5.2, shows the capital stock to GDP is equal to about 4 in 2001, and increases to 4.65 by year 2031. From 2001 to 2011, capital per worker increases by a factor of 1.44, and from 2001 to 2031 it increases by a factor of 2.44. At the half-way point to long-run equilibrium, GDP per worker increases by 39 percent, and by a factor of 2.09 by 2031. The difference between these results and the empirical example of Chapter 4 suggests adding intermediate inputs dampens slightly the predicted rate of transition growth in GDP.

The trends in total income and its components remain similar to the corresponding trends in the empirical example of Chapter 4. The ratio of labor income to total income, and of capital income to total income remains roughly equal throughout transition at about 49 percent for each factor – see Table 5.2. Returns to land show a non-monotonic path as in the previous example, with the value-added by land falling from 1.4 percent of total income initially to 0.73 percent by 2031. Expenditure per worker and the ratio of saving to GDP are only marginally smaller than those found in the preceding example. Although not shown here, the evolution of sector value shares in GDP follows the same pattern, except agriculture's value share in GDP is marginally higher while the industrial sector's share is marginally lower.

Tables 5.3, 5.4, and 5.5 present the results of a growth accounting exercise using the supply functions (5.11), (5.12) and (5.13). Compare the entries in Tables 5.3, 5.4, and 5.5, with the corresponding entries in Tables 4.4, 4.5, and 4.6. Although they converge to the same rate of growth in the long-run, growth in industrial and service sector gross output is slower with the addition of intermediate inputs. While these differences are small,

Table 5.2 Factor income and expenditure in millions of 2001 Turkish Lira

Year	GDP per worker	Capital per worker	Wage income per worker	Capital earnings per worker	Land rental income per worker	Expenditure per worker	Saving to GDP
2001	6522	25896	3262	3171	88	4003	0.39
2006	7549	31506	3759	3706	85	4689	0.38
2011	8621	37298	4280	4256	84	5405	0.37
2016	9749	43311	4832	4832	85	6155	0.37
2021	10947	49598	5419	5439	88	6949	0.37
2026	12229	56224	6049	6087	93	7795	0.36
2031	13608	63258	6727	6782	99	8701	0.36

Source: Model results

they compound over time. The rate of output growth in 2006 of the industrial sector in the model with intermediate inputs is smaller by a factor of 0.96 than its rate of growth in the previous example. By 2031, this difference narrows to a factor of 0.98. Corresponding values for the service sector are larger but less than unity. For the case of agriculture the growth in gross output is higher with intermediate inputs than without. In 2006, the ratio of growth in agricultural output with intermediate inputs to growth in agricultural output without intermediate inputs is 1.74, declining to 1.07 by 2031. Consequently, the shares in GDP of the industrial and service sector value-added are marginally lower in the model with intermediate inputs as compared to the corresponding value shares in Chapter 4, and higher for the case of agriculture.

Table 5.3 Growth in industry output and factor contributions

		Contributions to growth		
Year	Growth in gross output	Value added service price	Capital stock	Effective labor
2001	0.0625	−0.1678	0.4757	−0.2455
2006	0.0518	−0.1171	0.4003	−0.2314
2011	0.0458	−0.0843	0.3538	−0.2237
2016	0.0421	−0.0618	0.3231	−0.2191
2021	0.0397	−0.0459	0.3018	−0.2162
2026	0.0380	−0.0343	0.2866	−0.2143
2031	0.0369	−0.0258	0.2756	−0.2130

Source: Model results

Over time, the negative terms of trade effect from increased intermediate factor costs is shown by the negative effect of the industrial (agricultural) sector's decreasing value-added price on the industrial (agricultural) sector's rate of growth in gross output (Tables 5.3 and 5.4). The negative value-added price effect is more pronounced on industrial output than on agricultural output, and results because the industrial sector uses a larger share of services as intermediate inputs than the agricultural sector. As with the empirical model of Chapter 4, an increasing service price has a positive effect on service sector output. The service good own price effect, however, is smaller than that in Chapter 4, as the direct effect of the price increase is reduced by the relatively high share of service sector input in its own gross output (see Table 5.1).

Table 5.4 Growth in service output and factor contributions

		Contributions to growth		
Year	Growth in gross output	Value added service price	Capital stock	Effective labor
2001	0.0483	0.1230	−0.2620	0.1873
2006	0.0441	0.0884	−0.2305	0.1862
2011	0.0413	0.0646	−0.2086	0.1853
2016	0.0393	0.0477	−0.1930	0.1846
2021	0.0378	0.0355	−0.1818	0.1841
2026	0.0368	0.0266	−0.1736	0.1837
2031	0.0360	0.0200	−0.1674	0.1834

Source: Model results

The effect of changes in the value-added price on wage and capital rental rates are given by the reduced form wage and capital rental rate equations (5.9) and (5.10). The change in these rates on growth in agricultural output are shown in Table 5.5. Compared to the results of Chapter 4, the wage rate rises at a modestly slower rate while the interest rate falls at a slightly slower rate, a result of the impact of the increased service price on value-added prices. These indirect service good price effects explain the smaller negative wage effects and smaller positive capital rental rate effects on growth in agricultural gross output. The fundamental Rybczynski-like effects which are discussed at length in Chapter 4 still prevail and tend to dominate the contributions to growth.

The consequence of these growth patterns is a trade reversal as in Chapter 4. In about 2024 the economy switches from exporting the agricultural good to exporting industrial goods. In the long-run the value of exports plus imports accounts for about 6 percent of GDP.

Table 5.5 Growth in agriculture output and factor contributions

		Contributions to Growth			
Year	Growth in gross output	Value added service price	Wage effect	Interest rate effect	Technical change
2001	0.0019	−0.0031	−0.0730	0.0444	0.0336
2006	0.0099	−0.0024	−0.0545	0.0332	0.0336
2011	0.0158	−0.0018	−0.0409	0.0249	0.0336
2016	0.0203	−0.0013	−0.0307	0.0187	0.0336
2021	0.0235	−0.0010	−0.0232	0.0141	0.0336
2026	0.0260	−0.0008	−0.0175	0.0107	0.0336
2031	0.0278	−0.0006	−0.0133	0.0081	0.0336

Source: Model results

5.5.3 *Multiplier effects of a technology shock*

One motivation for including intermediate inputs in the model is to better capture their importance of inter-sectoral linkages in the process of growth. These linkages become important since tariffs can multiply up when goods are traded multiple times during stages of production. Jones (2007) illustrates this effect by showing how a positive shock to the scale parameter of technology in a Solow economy can lead to larger multiplier effects if intermediate inputs are considered.

International competition among firms in traded goods is likely to lead to similar technologies across countries. However, we suggest this outcome is less likely in the case of non-traded goods, as the disciplining impact of international competition may not make its way into markets that face only domestic competition. In Chapter 8, we specify the empirical analog of the service sector technology with intermediate inputs as

$$Y_s = \min_{L_s, K_s, Y_{as}, Y_{ms}, Y_{ss}} \left\{ \Psi_s \left(\mathcal{A}\left(t\right) L_s\right)^{\alpha_s} \left(K_s\right)^{1-\alpha_s}, \frac{Y_{as}}{\sigma_{as}}, \frac{Y_{ms}}{\sigma_{ms}}, \frac{Y_{ss}}{\sigma_{ss}} \right\}$$

where Ψ_s is the scale parameter. The simulation performed here entails increasing Ψ_s in the initial period by 10 percent for both the model without intermediate inputs and the model with intermediate inputs. We then compare the steady-state values of each model with and without the technology shock. The results are reported in Table 5.6.

Consider first the impact of a 10 percent increase in Ψ_s on the empirical model of Chapter 4. For this simulation, the shock leads to a 1.3 percent increase in the steady-state level of GDP (see Table 5.6). The capital stock level, and the level of industrial and agricultural production, remain unchanged from that of the base model. Also, the level of resources employed by the industrial and agricultural sectors remain unchanged. In this case, the price of the home-good falls, and the service sector produces an increase in output in the long run that is exactly equal to the 10 percent shock to its scale parameter, and otherwise employs the same level of resources as in the base model without the shock.

The same shock applied to the model with intermediate inputs has a far larger impact on the modeled economy. Table 5.6 shows that a 10 percent increase in Ψ_s leads to a long-run equilibrium in which the capital stock is 8 percent higher than the long-run equilibrium given by the same model without the shock. Gross domestic product increases by 7 percent of its value in the base intermediate factor model. The industrial and service sectors experience an increase in gross output equal to 13.3 percent and 11.2 percent, respectively. Agriculture's gross output falls by 23 percent of its value in the base model. The annual rate of growth in GDP in 2006 is 3.7 percent higher than that in the base model, and remains higher throughout transition: Of course, both models converge to the same rate of growth in the long-run. Consequently, the half-way point to a higher long-run level of GDP than the base model is reached in a slightly shorter period of time.

Throughout transition, the price of the service good is about 5 percent below its value in the base model. In this case, the value-added price faced by the industrial sector is about 2.7 percent higher throughout transition than in the base model. The value-added price for agriculture is about 1.3 percent higher than in the base model. This results because the share of the service good in industrial gross output is larger than the share of the service good in agricultural gross output (see Table 5.1). Hence, at each t the shock to the service sector tends to lower the cost of producing the industrial good relative to the cost of producing the agricultural good. Being more productive, the service sector needs fewer resources to clear its output market, and does so at a lower output price. Resources also flow out of agriculture which, after ten years, causes the sector to be import competing.

The impact on factor shares in GDP is also affected. The decline in gross agricultural output is accompanied by a decline in the returns to land that, on an annual comparison basis, ranges from a decline of 14 percent of the value in the base model in 2006, to 20 percent of base model value by 2031. While wage income and capital income both increase relative to the base model, the rise in income from capital exceeds the base model by 5 percent in 2006, while wage income is 3 percent higher. These results occur because the marginally improved terms of trade associated with a lower service good price favors the relatively capital intensive industrial sector as compared to the more labor intensive agricultural sector.

The economy also becomes more outward oriented. In the long-run, the value of industrial sector exports increase by 137 percent of the value of industrial sector exports predicted by the

Table 5.6 Multiplier effects from a 10% shock to the service technology scale parameter

	Ratio of steady-state values: Simulation to base model	
	No intermediate input model	Intermediate input model
Capital Stock	1.000	1.081
GDP	1.013	1.070
Industry output	1.000	1.133
Agriculture output	1.000	0.770
Service output	1.100	1.112

Source: Model results

base model. The value of exports plus imports as a percent of GDP rise from a base model prediction of 6 percent in the long-run to 12 percent of GDP in the intermediate factor simulation. Thus, as Yang (2008) finds from a study of the sources of export-led growth of 71 countries, the non-tradable sector may generate high economic growth together with high export growth.

5.6 Conclusions

This chapter sought to extend the highly stylized model of Chapter 4 in three ways. First we introduced intermediate inputs. We then modeled capital stock in a way that accounts for a composite of outputs from all sectors of the economy. Lastly, we showed how to accommodate government expenditure and sources of government revenue. Intermediate inputs allow for the examination of vertical market structures and to assess the importance of changes in a country's internal terms of trade in the process of economic growth.

Modeling capital as a composite of industrial, agricultural, and service sector output matches more closely actual input-output data. We defined the composite as the least cost combination of outputs. Such a definition permits the use of Shephard's lemma to decompose the composite into its specific components. A complication that emerges from the creation of composite capital is the price of capital becomes endogenous, and changes over time in proportion to the change in the price of the home-good. This complication requires the statement of a no arbitrage condition that links the land price to the rental rate firms pay for capital and the rental rate households earn on their capital asset.

The last extension considers government consumption and taxes. This modification is essential for policy analysis concerned with the economic growth effects of policy instruments such as indirect production taxes, taxes and subsidies on foreign trade, and government expenditures in excess of fiscal revenues. This extension also allows for the separation of private consumption from government consumption. Since government tends to consume mostly home-goods, household expenditure shares better reflect their actual shares which otherwise appear relatively large for the home-good and consequently small for other goods. This distinction is particularly important if households are presumed to hold Stone-Geary preferences.

The empirical example considered the intermediate factor extension of the model. We fit the model to the same data and parameters used in the previous chapters, and showed how the model's predictions departed from those of the previous chapter. Emphasis was placed on the internal terms of trade effect as capital deepening caused the price of the service good to rise, thus increasing the cost of this component of intermediate factor demand in total cost of production. Since the industrial sector employed a larger share of the service good in its gross value of production, an increase in the price of the service good, all else constant, was shown to have larger negative terms of trade effects on the industrial sector than on agriculture. We then performed a simulation in which we increased the scale parameter of the service sector production function by 10 percent. We motivated this simulation by suggesting that firms in the non-traded good sector are not under the pressure of international competition to employ the most efficient technology, and by the notion that export-led economic growth of many of the world's economies might be linked to technological efficiencies in the service sector that release resources to the traded good sectors. The results suggested that an increase in service sector productivity led to an increase in the rate of transition growth and to an increasing share of Turkey's GDP in international trade. For the case of Turkey, in the long-run, with the improved service technology more resources were pulled from both the service and the agricultural sectors and the economy exported more of the industrial good, while importing more of the agricultural good.

The task of the next chapter is to combine all of these extensions into a single model. We show that complexities arise that were not encountered when the extensions are modeled separately. The numerical example also focuses on a validation exercise in which the model's predictions, both forward and backward from the base year 2001, are compared to the data. This exercise also provides insights into the fit of the empirical models in this and the previous two chapters.

6

The Extended Three-Sector Model

Chapter 5 pointed out the importance of extending the basic three-sector model to accommodate (i) intermediate inputs of production, (ii) a stock of capital defined as a composite of various sector outputs, and (iii) government consumption and revenues. This chapter takes up the task of combining these extensions into a single model – the outcome of which is a model more suited to a broad array of policy analyses. Combining each of these features into a single model, however, has the cost of introducing a number of complications that make presenting and specifying the model more challenging.

To better facilitate the presentation of the extended model, and at the cost of some repetition of material in previous chapters, we begin with the household's problem, introduce government, and follow with the firms' problem. Then, following the same pattern as in previous chapters, we deal with the intra-temporal equilibrium, and conclude with the presentation of inter-temporal equilibrium. We also point out aspects of the model that were omitted in previous chapters, such as the no-arbitrage condition between capital and land, and determinants of the price of land. The last major section deals with fitting this more complex model to data. We also draw upon the empirical results to discuss additional features of the model that are otherwise difficult to obtain from the theory alone.

6.1 The model

As in Chapters 4 and 5, the environment is a small open economy that produces and consumes two traded goods – a manufacturing good, an agricultural good, and a non-traded home-good. Each of the three goods are produced by perfectly competitive

T.L. Roe et al., *Multisector Growth Models*, DOI 10.1007/978-0-387-77358-2_6,

firms. The manufactured and home-good sectors employ intermediate inputs and the primary factors of labor and capital, while agriculture employs intermediate inputs along with labor, capital and land. A share of all three goods is allocated to final household and government consumption, reinvested to increase the economy's stock of capital, and employed as intermediate inputs. Any surplus or deficit of the manufacturing and agricultural good is traded internationally. Labor services are not traded internationally, and there is no foreign ownership of capital. Government spends a constant share of GDP, and the share of government spending across goods is constant over time. These expenditures are balanced by indirect taxes on production, taxes on foreign trade and lump-sum transfers from households.

6.1.1 Households

As in the previous chapters, households are treated as an infinitely lived dynasty. The population of each household grows at a constant positive rate n. We assume a constant inter-temporal elasticity of substitution felicity function defined over a composite index q per worker. The representative household maximizes the discounted present value of future flows of utility given by

$$\int_0^\infty \frac{q(t)^{1-\theta}-1}{1-\theta} e^{(n-\rho)t} dt \tag{6.1}$$

where $1/\theta$ is the elasticity of inter temporal substitution, ρ is the rate of time preference, and we assume $\theta > 0$, $\rho > n$. Households receive income by providing the services of labor, capital and land in exchange for factor payments w, r, and π per worker, respectively, and pay lump-sum taxes T_{gov} to government. Capital and loans are assumed to be perfect substitutes. Unspent income accumulates as an asset for future consumption. The flow budget constraint of the representative household, expressed in per worker terms, is given by

$$\dot{p}_k k + p_k \dot{k} = w + (r-n)\, p_k k + \pi H + T_{gov} - \epsilon \tag{6.2}$$

where p_k and k are the price of capital and the stock of capital per worker, respectively, while

$$\epsilon = \mathcal{E}\left(p_a, p_m, p_s\right) q \tag{6.3}$$

is expenditure per worker.

As we have assumed throughout the text, ϵ is the minimum cost of achieving per worker composite consumption q given preferences $u\left(\cdot\right)$ and prices $\left(p_a, p_m, p_s\right)$. For convenience, we continue to assume constant expenditure shares expressed by λ_j, $j = a, m, s$, and the domestic price of manufactures, p_m, is treated as the numeraire price.

Notice that replacing r in the budget constraint by the no-arbitrage condition (5.39), Equation (6.2) becomes

$$\dot{k} = \frac{1}{p_k}\left(w + r^k k + \pi H + T_{gov} - \epsilon\right) - k\left(\delta + n\right) \tag{6.4}$$

Substituting (6.3) into (6.4) and solving the usual Hamiltonian problem maximizing (6.1) subject to (6.4), we obtain the Euler condition

$$\frac{\dot{\epsilon}}{\epsilon} = \frac{1}{\theta}\left(\frac{r^k}{p_k} - \rho - \delta - \left(1-\theta\right)\lambda_s \frac{\dot{p}_s}{p_s} + \frac{\dot{p}_k}{p_k}\right) \tag{6.5}$$

which is identical to that found in Chapter 5 for the case of composite capital. To rule out a Ponzi scheme, we require the transversality condition

$$\lim_{t\to\infty} k\left(t\right) e^{-\int_0^t \left(r(s) - n - \frac{\dot{p}_k}{p_k}\right) ds} = 0$$

However, our assumption that $\rho > n$, and the steady-state condition discussed later assures that this condition holds in the limit.

We next discuss the notion of composite capital, as laid out in the previous chapter, which serves to define the price of capital p_k, and allows us to replace the $\dot{p}_k/p_k$ term in (6.5) by an expression that depends on $\dot{p}_s/p_s$. We then focus on the no-arbitrage condition to show the link between the price of capital and the price of land.

Composite capital

As in the previous chapter, composite capital is modeled as combining in a least cost manner the incremental output y_{mk}, y_{ak} and y_{sk} of sector j per worker allocated to form composite capital. Equation (5.45) leads to the definition of the price of capital p_k as $c^k(p_a, p_m, p_s)$ where the function $c^k(\cdot)$ is homogeneous of degree one and non-decreasing in prices. Define

$$\tilde{c}^k(p_s) \equiv c^k(p_a, p_m, p_s)$$

Then, the total cost of an increment of capital per worker is equal to

$$\tilde{c}^k(p_s)\left(\dot{k} + k(\delta + n)\right) = p_k\left(\dot{k} + k(\delta + n)\right)$$

To derive the amount of sector j output required to produce one unit of composite capital, invoke Shepard's lemma. Doing so yields, in effective worker terms, the amount of sector j output used to produce composite capital

$$\hat{y}_{jk} = \frac{\lambda_{jk}}{p_j}\left(\widehat{gdp} + \hat{T}_{gov} - \hat{\epsilon}\right) \tag{6.6}$$

The parameter λ_{jk} is the cost share of Y_{jk} in the total cost of producing a unit of composite capital, $\hat{\epsilon} = \epsilon e^{-xt}$ is the minimum cost of achieving composite utility level $\hat{q}$ and

$$\widehat{gdp} = \hat{w} + r^k\hat{k} + \boldsymbol{\pi}^a\left(p_{va}, \hat{w}, r^k\right)H \tag{6.7}$$

is value-added per effective worker.[1] The presence of $\hat{T}_{gov} \equiv T_{gov}e^{-xt}$ is shown below to create a difficulty when characterizing the intra-temporal equilibrium.

Since traded good prices are fixed, the change in the price of capital results from the change in the price of the home-good. Consequently, by Equation (5.46) it follows that

[1] Here we use the fact that $\frac{\partial c^k(\cdot)}{\partial p_j}\left(\dot{k} + k(\delta + n)\right) = \frac{\lambda_{jk}p_k}{p_j}\left(\frac{1}{p_k}(w + r^k k + \pi H + T_{gov} - \epsilon)\right)$ and, from the budget constraint, $\dot{k} + k(\delta + n) = \frac{1}{p_k}(w + r^k k + \pi H + T_{gov} - \epsilon)$.

$\dot{p}_k/p_k = \lambda_{sk}\dot{p}_s/p_s$. The Euler condition can now be stated in effective worker terms as

$$\frac{\dot{\hat{\epsilon}}}{\hat{\epsilon}} = \frac{1}{\theta}\left(\frac{r^k}{\tilde{c}^k(p_s)} - \rho - \delta - \theta x + (\lambda_{sk} - (1-\theta)\lambda_s)\frac{\dot{p}_s}{p_s}\right) \quad (6.8)$$

The price of land

To simplify the land price derivation that follows, express total assets, A, in per worker terms as

$$a = k + p_h H$$

where H is the quantity of land and $p_h = p_L/p_k$ is the price of land per worker, p_L, relative to the unit price of capital p_k. Thus, the value of total assets per worker is

$$p_k a = p_k k + p_k p_h H$$

This definition of assets leads to the budget constraint

$$\dot{p}_k a + p_k \dot{a} = w + (r-n)\,p_k a + T_{gov} - \epsilon$$

Substituting $a = k + p_h H$ and $\dot{a} = \dot{k} + \dot{p}_h H$ into the above expression gives

$$\dot{p}_k k + \dot{p}_k p_h H + p_k \dot{k} + p_k \dot{p}_h H = w + (r-n)\,p_k k + (r-n) \\ p_k p_h H + T_{gov} - \epsilon \quad (6.9)$$

For budget constraint (6.2) to be equivalent to (6.9), it must be the case that

$$r = \frac{\dot{p}_k}{p_k} + \frac{\dot{p}_h}{p_h} + \frac{\pi/p_k}{p_h} + n \quad (6.10)$$

Expression (6.10) is the no-arbitrage condition between the rate of return to a unit of composite capital k, and a unit of land. Substituting $p_h = p_L/p_k$ and $\dot{p}_h/p_h = (\dot{p}_L/p_L - \dot{p}_k/p_k)$ into (6.10) and simplifying gives

$$r = \frac{\dot{p}_L}{p_L} + \frac{\pi}{p_L} + n = \frac{r^k}{p_k} - \delta + \frac{\dot{p}_k}{p_k}$$

This result is the same as (4.4) except that the price of land here is expressed in per worker terms.

Using (6.10) to solve the differential equation in p_h and the transversality condition reveals the time t normalized price of land is the discounted present value of the stream of rents π per worker

$$p_h(t) = \int_t^\infty e^{-\int_t^\tau \left(r(v)-n-\frac{\dot{p}_k}{p_k}\right)dv} \frac{\pi(\tau)}{p_k(\tau)} d\tau \qquad (6.11)$$

A similar result was first obtained by Nichols (1970) for a single sector model in which output was expressed as a function of capital, labor and land. He showed that the capital gains from an increase in the price of land over time can be a substitute of savings, thus lowering growth in a country's stock of capital.

6.1.2 Government

We employ the same passive role of government in the economy as the previous chapter. Government expenditure ϵ_g is a constant share λ_g of total primary factor payments (5.57), and λ_j^g is the share of government expenditures spent on the jth good. The government revenue equation (5.61) is

$$G_{rev} = \sum_{j=a,m} p_j^w \tau_j z_j + \sum_{j=a,m,s} p_j \tau_j^I y_j \qquad (6.12)$$

where net foreign trade, z_j, for the manufacturing good and agricultural good are

$$z_m \equiv y_m(1-\sigma_{mm}) - \sigma_{ma}y_a - \sigma_{ms}y_s - y_{mk} - q_m^g - q_m$$

and

$$z_a \equiv y_a(1-\sigma_{aa}) - \sigma_{am}y_m - \sigma_{as}y_s - y_{ak} - q_a^g - q_a$$

respectively. The revenue per worker from trade taxes or subsidies is the product of the world price p_j^w, the ad valorem tax rate τ_j, and net trade z_j.

We also include in government revenue, taxes (or subsidies) on production that are not assigned to a particular factor. We refer to this source of revenue as an indirect tax levied on the gross value of production. Revenue from indirect taxes on the value of output per worker is the product of the indirect rate τ_j^I, the domestic price p_j and gross output y_j. Other instruments could also be considered, such as a tax on wages or taxes on the sale of final goods to households.

Consumers face domestic prices for the manufactured and agricultural goods given by

$$p_j = p_j^w (1 + \tau_j), \quad j = a, m$$

Thus, consumer prices of the two traded goods only differ from world prices by the level of trade taxes or subsidies. Because of indirect taxes, producers face prices

$$p_j \left(1 - \tau_j^I\right), \quad j = a, m, s$$

To balance government expenditures with revenues, lump sum transfers are levied on households. The transfer per worker at each t is simply the difference between government revenues and expenditures

$$T_{gov} = G_{rev} - \epsilon_g \tag{6.13}$$

As one might suspect, this equation is more complex than the one in the previous chapter. As shown below, (6.13) creates a complication in the characterization of equilibrium because T_{gov} appears in the budget constraint (6.4) and in the equation for composite capital (6.6).

6.1.3 Firms

The technology of firms is given by (5.1) and (5.2) of the previous chapter. The unit cost function per effective worker for manufacturing and home-good production are the same as (5.3) except we now account for indirect taxes on production Thus, the minimum unit cost of producing the manufacturing and the home-good is given by

$$C^j\left(\hat{w}, r^k\right) + \sum_{i=a,m,s} \sigma_{ij} p_i + \tau_j^I p_j, \quad j = m, s$$

The rent to agricultural value-added per effective worker is given by

$$\boldsymbol{\pi}^a \left(p_{va}, \hat{w}, r^k\right) H$$

where we remind the reader the rate of land augmentation is given by $\boldsymbol{B}\ (t) = \mathcal{A}(t) L(t)$ – a sustainability condition that yields differential equations that are autonomous. We also redefine the value-added price to producers of the three goods to include indirect taxes,

$$p_{vj} = p_j \left(1 - \tau_j^I\right) - \sum_{i=a,m,s} \sigma_{ij} p_i$$

It is now clear that trade taxes have a direct and an indirect effect on producer incentives. This can be seen more clearly by comparing the value-added price

$$p_{vj} = p_j \left(1 - \tau_j^I\right) - \sum_{i=a,m} \sigma_{ij} p_i - \sigma_{sj} p_s$$

faced by producers of the two traded goods with the undistored value-added price

$$p_{vj}^w = p_j^w \left(1 - \tau_j^I\right) - \sum_{i=a,m} \sigma_{ij} p_i^w - \sigma_{sj} p_s^*$$

that would prevail in the absence of trade taxes, where p_s^* is the equilibrium price of the home-good that prevails in the undistorted case. The commonly used nominal rate of protection measure, $\tau_j = \left(p_j - p_j^w\right)/p_j^w$ amounts to a partial estimate of the direct effect.[2] The indirect effect is transmitted through the cost of intermediate inputs. This effect is proportional to the input-output coefficient σ_{ij}, $i, j = a, m$ for traded goods employed as intermediate inputs plus the intermediate input effect of the home-good. This effect is proportional to the product of the input-output coefficient σ_{sj} and the extent to which trade

[2] This measure must be interpreted with caution since the supply function is typically viewed as being homogeneous of degree zero in output prices so that relative prices affect incentives.

taxes affect the evolution of the home-good price p_s relative to the home-good price p_s^* that would prevail in the absence of trade taxes. The home-good effect can be positive or negative depending upon whether p_s tends to converge from below or above its long-run value. An empirical estimate of the total effect of trade taxes on incentives for producing the jth traded good is given by

$$\frac{p_{vj}/p_{vs} - p_{vj}^w/p_{vs}^*}{p_{vj}^w/p_{vs}^*}, \quad j = a, m \tag{6.14}$$

where a positive value suggests protection, a negative value disprotection.

As shown in Table 5.1, intermediate input share in the value of gross output can be relatively large, and vary by sector which causes the indirect effects to impact various sector unequally. Thus, price distortions caused by tariffs in this more detailed model can have a greater effect on the modeled economy than they do in the previous models.

6.1.4 Intra-temporal equilibrium

Given the sequence $\left\{\hat{k}(t), \hat{\epsilon}(t)\right\}_{t\in[0,\infty)}$ of capital stock and expenditure pairs, an intra-temporal equilibrium is characterized by the five-tuple sequence of positive values

$$\{\hat{w}(t), r^k(t), \hat{y}_m(t), \hat{y}_s(t), p_s(t)\}_{t\in[0,\infty)}$$

that satisfy an equal number of equations for each t. Four of these, by now familiar equations, are the following:

- zero profits in sectors m, s

$$C^j\left(\hat{w}, r^k\right) = p_{vj}, \quad j = m, s; \quad p_{vj} = p_j - \sum_{i=a,m,s} a_{ij}p_i - \tau_j^I p_j \tag{6.15}$$

- labor market clearing

$$\sum_{j=m,s} C_{\hat{w}}^j\left(\hat{w}, r^k\right)\hat{y}_j - \boldsymbol{\pi}_{\hat{w}}^a\left(p_{va}, \hat{w}, r^k\right)H = 1 \tag{6.16}$$

- and capital market clearing

$$\sum_{j=m,s} C^{j}_{rk}\left(\hat{w}, r^{k}\right)\hat{y}_j - \boldsymbol{\pi}^{a}_{rk}\left(p_{va}, \hat{w}, r^{k}\right) H = \hat{k} \tag{6.17}$$

The home-good market clearing condition,

$$\frac{\partial \mathcal{E}\left(p_a, p_s\right)\hat{q}}{\partial p_s} = \frac{\lambda_s \hat{\epsilon}}{p_s} = \hat{y}_s\left(1 - \sigma_{ss}\right) - \sigma_{sm}\hat{y}_m - \sigma_{sa}\hat{y}_a - \hat{y}_{sk} - \hat{q}^{g}_{s} \tag{6.18}$$

is used to derive the fifth equation needed to complete this system. Our task is to replace the three terms $\hat{y}_a$, $\hat{y}_{sk}$ and $\hat{q}^g_s$ with their reduced form expressions that contain, at most, members of the following endogenous variables as arguments, $\hat{w}$, r^k, p_s, and $\hat{k}$ and $\hat{\epsilon}$.

The reduced forms for $\hat{y}_a$ and $\hat{q}^g_s$ are easily dealt with. Agricultural supply, as in the prior chapter, is

$$\hat{y}_a = \boldsymbol{\pi}^{a}_{p_{va}}\left(p_{va}, \hat{w}, r^{k}\right) H \tag{6.19}$$

Government consumption of the home-good is

$$\hat{q}^g_s = \frac{\lambda^g_s \lambda_g \widehat{gdp}}{p_s} \tag{6.20}$$

where $\widehat{gdp}$ is given by (6.7). If home-good demand, $\hat{y}_{sk}$, used to produce composite capital did not appear in (6.18), the number of variables and equations would be equal.

Home-good demand allocated to composite capital can be expressed as

$$\hat{y}_{sk} = \frac{\lambda_{sk}}{p_s}\left(\widehat{gdp} + \hat{T}_{gov} - \hat{\epsilon}\right) \tag{6.21}$$

where the term in brackets equals $\dot{\hat{k}} + \hat{k}\left(x + \delta + n\right)$. The presence of government transfers $\hat{T}_{gov}$ creates a difficulty that did not arise in the previous chapter.

If we substitute (6.21) for $\hat{y}_{sk}$ in (6.18), the resulting system (6.15), (6.16), (6.17), and (6.18) for given $\left\{\hat{k}\left(t\right), \hat{\epsilon}\left(t\right)\right\}_{\tau\in[0,\infty)}$, leaves us with the extra endogenous lump-sum transfer term $\hat{T}_{gov}$. The next task is to replace $\hat{T}_{gov}$ in (6.21) in order to square the system of Equations (6.15), (6.16), (6.17), and (6.18).

Express (6.12) in effective worker terms and use it to substitute for G_{rev} in (6.13); the resulting transfer equation becomes

$$\hat{T}_{gov} = p_m^w \tau_m \hat{z}_m + p_a^w \tau_a \hat{z}_a - \tag{6.22}$$

$$\left(\frac{\tau_m}{(1+\tau_m)}\lambda_{mk} + \frac{\tau_a}{(1+\tau_a)}\lambda_{ak}\right)\hat{T}_{gov} + \sum_{j=a,m,s} p_j \tau_j^I \hat{y}_j - \hat{\epsilon}_g$$

where we re-express net trade z_j as

$$\hat{z}_j \equiv \hat{y}_j\left(1-\sigma_{jj}\right) - \sigma_{ji}\hat{y}_i - \sigma_{js}\hat{y}_s -$$

$$\left(\frac{\lambda_j^g \lambda_g}{p_j} + \frac{\lambda_{jk}}{p_j}\right)\widehat{gdp} + \left(\frac{\lambda_{jk}}{p_j} - \frac{\lambda_j}{p_j}\right)\hat{\epsilon},\ i \neq j = a, m \tag{6.23}$$

To derive (6.23), use $\hat{q}_j^g = \lambda_j^g \lambda_g \widehat{gdp}/p_j$, and $\hat{q}_j = \lambda_j \hat{\epsilon}/p_j$. Separate the expression for $\hat{y}_{jk}$, Equation (6.6), into the part associated with $(\lambda_{jk}/p_j)\left(\widehat{gdp} - \hat{\epsilon}\right)$ and the part associated with $(\lambda_{jk}/p_j)\,\hat{T}_{gov}$ since we need to solve for $\hat{T}_{gov}$ below. The latter term simplifies to

$$\left(p_m^w \tau_m \frac{\lambda_{mk}}{p_m} + p_a^w \tau_a \frac{\lambda_{ak}}{p_a}\right)\hat{T}_{gov}$$

Replacing the domestic price p_j by $p_j^w\left(1+\tau_j\right)$ in this expression leads to

$$\frac{\tau_m}{(1+\tau_m)}\lambda_{mk} + \frac{\tau_a}{(1+\tau_a)}\lambda_{ak}$$

which appears in (6.22).

Expressing (6.22) in terms of $\hat{T}_{gov}$ gives

$$\hat{T}_{gov} = \frac{p_m^w \tau_m \hat{z}_m + p_a^w \tau_a \hat{z}_a + \sum_{j=a,m,s} p_j \tau_j^I \hat{y}_j - \lambda_g \widehat{gdp}}{\left(1 + \frac{\tau_m}{(1+\tau_m)}\lambda_{mk} + \frac{\tau_a}{(1+\tau_a)}\lambda_{ak}\right)} \tag{6.24}$$

The demand for the home-good allocated to composite capital, Equation (6.21), can now be expressed as

$$\frac{\lambda_{sk}}{p_s}\left(\widehat{gdp}+\frac{p_m^w\tau_m\hat{z}_m+p_a^w\tau_a\hat{z}_a+\sum_{j=a,m,s}p_j\tau_j^I\hat{y}_j-\lambda_g\widehat{gdp}}{\left(1+\frac{\tau_m}{(1+\tau_m)}\lambda_{mk}+\frac{\tau_a}{(1+\tau_a)}\lambda_{ak}\right)}-\hat{\epsilon}\right)$$

This equation contains endogenous variables that are common to the system (6.15), (6.16), (6.17), and (6.18). To maintain the pattern of characterization of the previous chapters, we substitute (6.24) into the composite capital demand equation (6.21), and then substitute this result into the home-good market clearing equation (6.18) for $\hat{y}_{sk}$. Replace $\hat{y}_a$ in (6.18) with (6.19) and $\hat{q}_{gs}$ with (6.20) which together completes the system (6.15), (6.16), (6.17), and (6.18).

6.1.5 Reducing the dimensionality of the system

As in the previous chapters, the zero profit conditions are used to express factor payments as a function of the value-added prices of the manufacturing and the home-good

$$\hat{w}=\tilde{w}(p_s)\equiv W(p_{vm},p_{vs}) \tag{6.25}$$

$$r^k=\tilde{r}(p_s)\equiv R(p_{vm},p_{vs}) \tag{6.26}$$

Substitute Equations (6.25) and (6.26) into expressions (6.16) and (6.17), and solve the two factor market clearing conditions to obtain the supply functions for manufacturing

$$\hat{y}_m=\tilde{y}^m\left(p_s,\hat{k}\right)\equiv y^m\left(p_{va},p_{vm},p_{vs},\hat{k},H\right) \tag{6.27}$$

the home-good

$$\hat{y}_s=\tilde{y}^s\left(p_s,\hat{k}\right)\equiv y^s\left(p_{va},p_{vm},p_{vs},\hat{k},H\right) \tag{6.28}$$

and, using (6.19), for agriculture,

$$\hat{y}_a=\tilde{y}^a(p_s)\equiv \boldsymbol{\pi}_{p_{va}}^a\left(p_{va},\tilde{w}(p_s),\tilde{r}(p_s)\right)H \tag{6.29}$$

Since transfers $\hat{T}_{gov}$ appear in the budget constraint, a key differential equation used later, we next express transfers in terms of the endogenous variables p_s, $\hat{k}$, and $\hat{\epsilon}$ which are arguments common to the reduced form equations constituting the intra-temporal system (6.15), (6.16), (6.17), and (6.18). Consider first (6.23) for $j = m$. Substituting the supply equations (6.27), (6.28), and (6.29) into (6.23) yields

$$\tilde{z}_m\left(p_s, \hat{k}, \hat{\epsilon}\right) =$$
$$\tilde{y}^m\left(p_s, \hat{k}\right)(1 - \sigma_{mm}) - \sigma_{ma}\tilde{y}^a(p_s) - \sigma_{ms}\tilde{y}^s\left(p_s, \hat{k}\right)$$
$$-\left(\frac{\lambda_m^g \lambda_g}{p_m} + \frac{\lambda_{mk}}{p_m}\right)\tilde{G}\left(p_s, \hat{k}\right) + \left(\frac{\lambda_{mk}}{p_m} - \frac{\lambda_m}{p_m}\right)\hat{\epsilon}$$

where

$$\tilde{G}\left(p_s, \hat{k}\right) = \tilde{w}(p_s) + \tilde{r}(p_s)\hat{k} + \tilde{\pi}(p_s) H$$

and

$$\tilde{\pi}(p_s) = \boldsymbol{\pi}^a(p_{va}, \tilde{w}(p_s)\tilde{r}(p_s))$$

Notice that $\tilde{z}_m(\cdot)$ is a function of the three endogenous variables p_s, $\hat{k}$ and $\hat{\epsilon}$. The same type of result is obtained for the case of agriculture, which we denote by $\tilde{z}_a\left(p_s, \hat{k}, \hat{\epsilon}\right)$. Then, substituting $\tilde{z}_m(\cdot)$, $\tilde{z}_a(\cdot)$, and Equations (6.27), (6.28), and (6.29) into (6.24) yields a transfer equation that as a function of p_s, $\hat{k}$, and $\hat{\epsilon}$:

$$\tilde{T}^g\left(p_s, \hat{k}, \hat{\epsilon}\right) =$$
$$\frac{p_m^w \tau_m \tilde{z}_m\left(p_s, \hat{k}, \hat{\epsilon}\right) + p_a^w \tau_a \tilde{z}_a\left(p_s, \hat{k}, \hat{\epsilon}\right) + p_m \tau_m^I \tilde{y}^m\left(p_s, \hat{k}\right)}{1 + \frac{\tau_m \lambda_{mk}}{1+\tau_m} + \frac{\tau_a \lambda_{ak}}{1+\tau_a}}$$
$$+\frac{p_a \tau_a^I \tilde{y}^a(p_s) + p_s \tau_s^I \tilde{y}^s\left(p_s, \hat{k}\right) - \lambda_g \tilde{G}\left(p_s, \hat{k}\right)}{1 + \frac{\tau_m \lambda_{mk}}{1+\tau_m} + \frac{\tau_a \lambda_{ak}}{1+\tau_a}}$$

The reader can verify the above equation is linear in $\hat{k}$ and $\hat{\epsilon}$.

The right hand side of the budget constraint can now be restated as a function of three endogenous variables, $\hat{k}$, p_s and $\hat{\epsilon}$.

$$\dot{\hat{k}} = \frac{1}{\tilde{c}^k(p_s)}\left[\tilde{w}(p_s) + \tilde{r}(p_s)k + \tilde{\boldsymbol{\pi}}^a(p_s)H + \tilde{T}^g\left(p_s, \hat{k}, \hat{\epsilon}\right)\right] - \frac{1}{\tilde{c}^k(p_s)}\hat{\epsilon} - \hat{k}(x+\delta+n) \tag{6.30}$$

In the previous chapters, the models are reduced to two differential equations. At this stage of reduction, it appears that the system can be expressed by three differential equations in p_s, $\hat{k}$ and $\hat{\epsilon}$. To eliminate one of the differential equations and adhere to the pattern of the previous models, use the home-good market clearing condition (6.18) to replace $\hat{\epsilon}$ in the budget constraint. To do this, first, use (6.21) to re-state the home-good demand allocated to the production of composite capital as

$$\hat{y}_{sk} = \tilde{y}^{sk}\left(p_s, \tilde{k}, \hat{\epsilon}\right) \equiv \frac{\lambda_{sk}}{p_s}\left(\tilde{G}\left(p_s, \hat{k}\right) + \tilde{T}^g\left(p_s, \hat{k}, \hat{\epsilon}\right) - \hat{\epsilon}\right)$$

and re-state the government's home-good consumption (6.20) as

$$\hat{c}_{gs} = \tilde{c}^{gs}\left(p_s, \hat{k}\right) \equiv \frac{\lambda_{gs}\lambda_g\tilde{G}\left(p_s, \hat{k}\right)}{p_s}$$

Proceed by plugging these two expressions and the supply functions (6.27), (6.28), and (6.29) into the home-good market clearing condition (6.18). After some simplification, we obtain,

$$\lambda_s\hat{\epsilon} = p_s\tilde{y}^s\left(p_s, \hat{k}\right)(1-\sigma_{ss}) - p_s\sigma_{sm}\tilde{y}^m\left(p_s, \hat{k}\right) - p_s\sigma_{sa}\tilde{y}^a(p_s) - \lambda_{sk}\left(T^g\left(p_s, \hat{k}, \hat{\epsilon}\right) - \hat{\epsilon}\right) - (\lambda_{sk} + \lambda_{gs}\lambda_g)\tilde{G}\left(p_s, \hat{k}\right) \tag{6.31}$$

Given the assumptions on preferences and technologies assumed throughout the text, one can show that (6.31) is linear in expenditure $\hat{\epsilon}$. Using this equation to isolate expenditure per effective worker, express the result as

$$\hat{\epsilon} = \tilde{\epsilon}\left(p_s, \hat{k}\right) \tag{6.32}$$

Substitute (6.32) into the budget constraint (6.30), and refer to the result as

$$\dot{\hat{k}} = \tilde{K}\left(p_s, \hat{k}\right) \equiv$$

$$\frac{1}{\tilde{c}^k\left(p_s\right)}\left[\tilde{w}\left(p_s\right) + \tilde{r}\left(p_s\right)k + \tilde{\boldsymbol{\pi}}^a\left(p_s\right)H + \tilde{T}^g\left(p_s, \hat{k}, \tilde{\epsilon}\left(p_s, \hat{k}\right)\right)\right] - \tag{6.33}$$

$$\frac{1}{\tilde{c}^k\left(p_s\right)}\tilde{\epsilon}\left(p_s, \hat{k}\right) - \hat{k}\left(x + \delta + n\right)$$

We now proceed with a discussion of the inter-temporal equilibrium conditions.

6.1.6 Inter-temporal equilibrium

If a steady-state exists, the Euler condition (6.8) implies the root p_s^{ss} satisfying

$$\frac{\tilde{r}\left(p_s\right)}{\tilde{c}^k\left(p_s\right)} = \rho + \delta + \theta x \tag{6.34}$$

is the steady-state price of the home-good.[3] As in the preceding chapter, given p_s^{ss}, use expressions (6.25) and (6.26) to calculate the values of $\hat{w}^{ss}$ and $r^{k,ss}$. Substitute these values into Equation (6.33), and solve for the root $\hat{k}^{ss}$ that satisfies $\dot{\hat{k}} = 0$. Substitute $\hat{k}^{ss}$ and p_s^{ss} into (6.32) to get the steady-state level of expenditure per efficient worker, $\hat{\epsilon}^{ss}$. Finally, recover the remaining endogenous variables using the intra-temporal conditions.

The differential equations of this more detailed model are derived in the same manner as in the previous chapters, although the resulting expressions are more complex than those associated with the models of Chapters 3, 4, and 5. The differential equation for capital is of course (6.33). The differential equation for $\dot{p}_s$ is obtained from the home-good market clearing, equation (6.32). Time differentiate this equation to obtain

$$\dot{\hat{\epsilon}} = \tilde{\epsilon}_{p_s}\left(p_s, \hat{k}\right)\dot{p}_s + \tilde{\epsilon}_{\hat{k}}\left(p_s, \hat{k}\right)\dot{\hat{k}}$$

[3] Given constant returns to scale Cobb-Douglas technologies and an interior solution which we have assumed throughout, it can be shown from the zero profit conditions that this solution is single valued and unique.

Substitute (6.32) for $\hat{\epsilon}$ in the Euler condition,

$$\dot{\hat{\epsilon}} = \frac{\tilde{\epsilon}\left(p_s, \hat{k}\right)}{\theta}\left(\frac{r^k}{\tilde{c}^k\left(p_s\right)} - \delta - \rho - \theta x + \left(\lambda_{sk} - (1-\theta)\,\lambda_s\right)\frac{\dot{p}_s}{p_s}\right)$$

and then substitute this result for $\dot{\hat{\epsilon}}$ obtain

$$\frac{\tilde{\epsilon}\left(p_s, \hat{k}\right)}{\theta}\left(\frac{\tilde{r}\left(p_s\right)}{\tilde{c}^k\left(p_s\right)} - \delta - \rho - \theta x + \left(\lambda_{sk} - (1-\theta)\,\lambda_s\right)\frac{\dot{p}_s}{p_s}\right)$$

$$= \tilde{\epsilon}_{p_s}\left(p_s, \hat{k}\right)\dot{p}_s + \tilde{\epsilon}_{\hat{k}}\left(p_s, \hat{k}\right)\dot{\hat{k}}$$

Finally, substitute (6.33) for $\dot{\hat{k}}$ and solve for $\dot{p}_s$ to obtain the differential equation

$$\dot{p}_s = \frac{p_s\left(\tilde{\epsilon}\left(p_s, \hat{k}\right)\left(\frac{\tilde{r}(p_s)}{\tilde{c}^k(p_s)} - \delta - \rho - \theta x\right) - \theta\tilde{\epsilon}_{\hat{k}}\left(p_s, \hat{k}\right)\tilde{K}\left(p_s, \hat{k}\right)\right)}{\theta p_s \tilde{\epsilon}_{p_s}\left(p_s, \hat{k}\right) - \left(\lambda_{sk} - (1-\theta)\,\lambda_s\right)\tilde{\epsilon}\left(p_s, \hat{k}\right)} \tag{6.35}$$

If a steady-state $(p_s^{ss}, \hat{k}^{ss})$ exists, the numerator is zero. Notice the similarity of this result with that for the case of composite capital alone (5.56). For θ equal to unity, the basic form of the equations are identical. Of course, this equation is more complex owing to the addition of government and intermediate inputs.

6.2 Numerical analysis

In Chapters 3, 4, and 5, we linked the theoretical model of each chapter to its empirical analogue. The common thread for each empirical model was the data for Turkey. The results of each simulation revealed the empirical two-sector closed economy model yielded similar results for major economy aggregates over time, as the three-sector model, which in turn yielded similar results as the three-sector model with intermediate inputs.

The empirical model of this chapter includes each of the extensions discussed in Chapter 5, i.e., intermediate inputs, composite capital, and government. As with each of the prior empirical models, the empirical model of this chapter also generates similar results in terms of the transition path of GDP, capital stock per worker, sectoral composition, and factor payment rates.

Given that each empirical model performs similarly in predicting the macroeconomic aggregates of sectoral output, in this section we take a look at how well this model predicts aggregate and sectoral output for Turkey between 1995 and 2005. We first present the parameter values that are new to the extended model of this chapter. We then discuss some of the issues regarding model validation, and confront the model's forecasts of aggregate and sectoral output values to data over the 1995–2005 period. Finally, we discuss the numerical results of the model and contrast these results with a solution to the model in which we decrease by 20 percent the tariff rate protecting industrial sector firms.

6.2.1 Parameter estimation

New to this example are the simultaneous inclusion of household and government expenditure shares (λ_j and λ_j^g, respectively), the cost share of sector j output in the total cost of composite capital (λ_{jk}), and tax revenues, all of which are shown in Table 6.1. The parameters reported in Table 4.1 remain unchanged, as do the share of intermediate factor demands in sectoral gross output values reported in Table 5.1. The social accounting matrix from which the data for this example are taken appears in Chapter 8.

The GTAP data (2001) indicate government expenditure as a share of GDP was 24 percent, and over 90 percent of this expenditure was on service good consumption. Disaggregating total consumption into household consumption and government consumption increases household expenditure shares on agricultural and industrial consumption as compared to the corresponding consumption shares in each of the prior empirical examples. The GTAP data also show (i) industrial and service sector goods comprise nearly all of composite capital, and (ii) the bulk of government revenue comes from indirect production taxes on industrial and service sector firms. Taxes from foreign trade account for only 4.3 percent of total tax receipts. However, in the empirical model discussed below, this tax amounts to the equivalent of an ad-valorem tariff rate (τ_m) on net industrial imports of almost 31 percent. Most of the remaining tax receipts come from an assortment of other taxes linked to workers and households. We model these remaining revenue sources as lump-sum transfers from households that just balance the government's budget in each period – see Equation (6.22).

Table 6.1 Shares: expenditure, composite capital and tax revenues

	Expenditure Share		Share in	
	Household	Government	composite capital	tax revenue
Industry	$\lambda_m = 0.240$	$\lambda_m^g = 0.039$	$\lambda_{mk} = 0.474$	0.359
Agriculture	$\lambda_a = 0.231$	$\lambda_a^g = 0.026$	$\lambda_{ak} = 0.021$	0.001
Service	$\lambda_s = 0.529$	$\lambda_s^g = 0.935$	$\lambda_{sk} = 0.505$	0.225
Household				0.372
Trade				0.043

Source: Author estimates using WDI and GTAP data

6.2.2 Validation

An overview of the issues

The dearth of exercises to validate structural models has been the source of critiques concerning the value of such models in explaining and predicting economic events.[4] Kehoe (2003) suggests ex-post performance evaluations of applied general equilibrium models are essential if policy makers are to have confidence in the results produced by them. He suggests ex-post evaluations

[4] Many early practitioners of applied general equilibrium models dismissed the need for validation. Whalley (1988) suggests these models are not intended to forecast values but instead to provide useful insights to policy makers of the various market interactions that can only be provided by a structural model of an economy.

can help make applied general equilibrium analyses a scientific discipline in which there are well-defined puzzles with clear successes and failures for competing theories.

Clearly, general equilibrium models are prone to specification and estimation biases. For example, a specification presuming perfect foresight and altruistic behavior of households is almost surely violated in reality. Another violation is that in each model presented thus far, the rest of the world is modeled as though it is in long-run equilibrium – an assumption implied when assuming traded good prices are fixed and time invariant. Another set of issues relate to the methods used for estimating some parameters, like a country's initial capital stock $K(0)$ and productivity growth rate x, both of which tend to be sensitive to a number of growth accounting assumptions, and are almost always prone to considerable variation over time.[5] Values of other parameters, such as household expenditure shares, are unlikely to remain constant with growth in income, yet are poorly depicted by Stone-Geary or other non-homothetic preference structures that permit aggregation over households.

The models presented here take no account of stochastic events that shock an economy. An immediate consequence of this model structure is the trajectory of all endogenous variable predictions are best interpreted as attempts to capture growth trends embedded in the data, and to provide an explanation of the economic and structural forces underlying these trends. A slight modification of this interpretation is that the model's solution for period $t = 0$ predicts the actions of agents in that period, while the predicted values for future periods, $t \geq 1$ are plans. These plans, however, may not predict future outcomes of a real economy, as unanticipated events induce agents to revaluate their plans and re-optimize in future periods. In any of these cases, the validation measures should not penalize the model for what it is not intended to capture.

Thus, the modeler faces a number of trade-offs in model specification, parameter estimation, and identifying criteria for evaluating model success or failure. In his attempt to answer

[5] See for example Caselli (2005).

the question – "What is game theory trying to accomplish?" Nobel Laureate Robert Aumann (1985) advances the view that a model is a caricature, or metaphor, of an environment. Instead of asking whether the model is right or wrong, the more germane question is "how useful is it?". For our purposes, "usefulness" depends, in part, on the model's capacity to replicate the trajectory of some of the economy's key variables of interest. Another criteria is the model's usefulness in explaining various puzzles embedded in the economic growth and development process.

Validation measures and results

The validation problem has been addressed by others prior to Kehoe's contribution, e.g., see Watson (1993), and Hansen and Heckman (1996). At present, however, it appears no universally agreed upon methodology for confronting applied dynamic general equilibrium models to data exists. Li and Roe (2006) experimented with several methods in their application of a three-sector growth model to Taiwanese data.[6] They developed a time series confidence interval method using an ARIMA process to assess whether their model's sectoral aggregate forecasts fell within the confidence limits of the ARIMA forecasts. They also employed a bootstrapping method that allowed the placement of confidence intervals on the model forecasts. While these approaches show promise, here we only show the results from more conventional methods.

The empirical model in this chapter is fit to 2001 GTAP data for Turkey, and then solved to provide predictions from 1995 onwards. Typically, a structural model will generate solutions for a larger set of variables than are available in the form of time series data. This is certainly the case here, and below we compare our model's predictions of aggregate and sectoral GDP to data taken from the WDI. These predictions and their corresponding actual data are shown in Figure 6.1.

---

[6] Measures of the Taiwanese model's fit to the economy are superior to those reported here for Turkey, possibly because Taiwan did not experience the frequency and magnitude of cyclical behavior as Turkey.

The upper left-hand chart of Figure 6.1 shows the aggregate GDP as reported in the WDI, and the model's forecast of GDP in trillions of 2001 Turkish lira. The model predictions appear to move with the WDI data, but it tends to undershoot the actual time series. In this case, the major reason for undershooting is that the model is calibrated to the trough of one of Turkey's business cycles. Unreported simulations confirm that had the model been fit to data at the top of a business cycle, model predications would exceed observations for many of the within-sample years.

Other factors also affect the predictive ability of the model. One is the data needed to capture the inter-temporal equilibrium along the transition path is typically not available. That which is available, e.g., expenditure data, is not exactly consistent with our estimate of the capital stock and the rate of return to capital needed to satisfy the initial period solution to the Euler condition. Another factor is whether estimates of the structural parameters, such as ρ, θ, δ and the estimate of a country's stock of capital $K(0)$ should be treated as being independent of the level of model aggregation of the economy, as in the case of our examples[7].

The three remaining charts in Figure 6.1 present the WDI series for agricultural, industry, and service sector GDP and the model's prediction of these series. Each series is normalized by its base year 2001 value. We employ this normalization because the International Standard Industrial Classification (ISIC) that defines the sectors in the WDI are not the ideal classification needed for the model. For example, agriculture corresponds to ISIC divisions 1–5 in the WDI data. This division includes crop and animal agriculture, hunting and related service activities, forestry, fishing and mining of coal. Food products, however, are included in manufacturing. On the other hand, the GTAP data includes food products in agriculture, and coal in

[7] The model's fit to the data are sensitive to these parameters so that relatively small changes in their values can cause forecasts to trend away from data. Yet, these small changes are often within the subjective confidence limits of the researcher's belief as to their true values.

manufacturing. Thus, the ISIC 1–5 aggregation is not an ideal characterization of agricultural and food output for our model, as the additional subsectors in GTAP's agriculture cause the model's agricultural output to exceed that of the WDI by about 26 percent in the initial year, 2001. Similar adjustments are made to the service and industrial sectors which cause industrial sector GDP to be only 81 percent of industrial GDP in the WDI, while the service sector GDP is about 6 percent larger than service GDP in the WDI.

In spite of the differences between the WDI and the model's definition of sectoral GDP, the predicted values show a similar trend to the data, suggesting the model is capturing some of the fundamental structural features of the economy. The left panel of Figure 6.1 shows the ISIC 1–5 subsector aggregation of agriculture, while the other line is the model forecast for agriculture, which of course includes additional subsectors, normalized to base year 2001 values. As with aggregate GDP, 2001 is a year in which agricultural production is at a low point of a cycle. A similar argument applies to the industrial and service sectors.

Numerical measures of the model's forecast accuracy are reported in Table 6.2. Pearson's correlation coefficient provides a linear measure of the correlation between the data and the forecast, without accounting for differences in the level of the variables in the two series. As can be seen from Figure 6.1, the variation apparent in agriculture's GDP results in the lowest Pearson measure. Lin's (1989) concordance correlation measure is bounded between zero and unity, and accounts for discrepancies between the means of the two series.[8] The result of this measure confirms the larger forecasting error for agriculture. The mean absolute error is relatively low for the agricultural

[8] Concordance Correlation Coefficient measure is given by $2S_{12}/(S_1^2 + S_2^2 + (\bar{Y}_1 - \bar{Y}_2)^2)$, where S_{12} is the covariance of model and data, the S_j^2 are their respective variance and the $\bar{Y}_j$ are means, $j = GDP$, $GDP_{agriculture}$, $GDP_{manufacturing}$, and $GDP_{service}$.

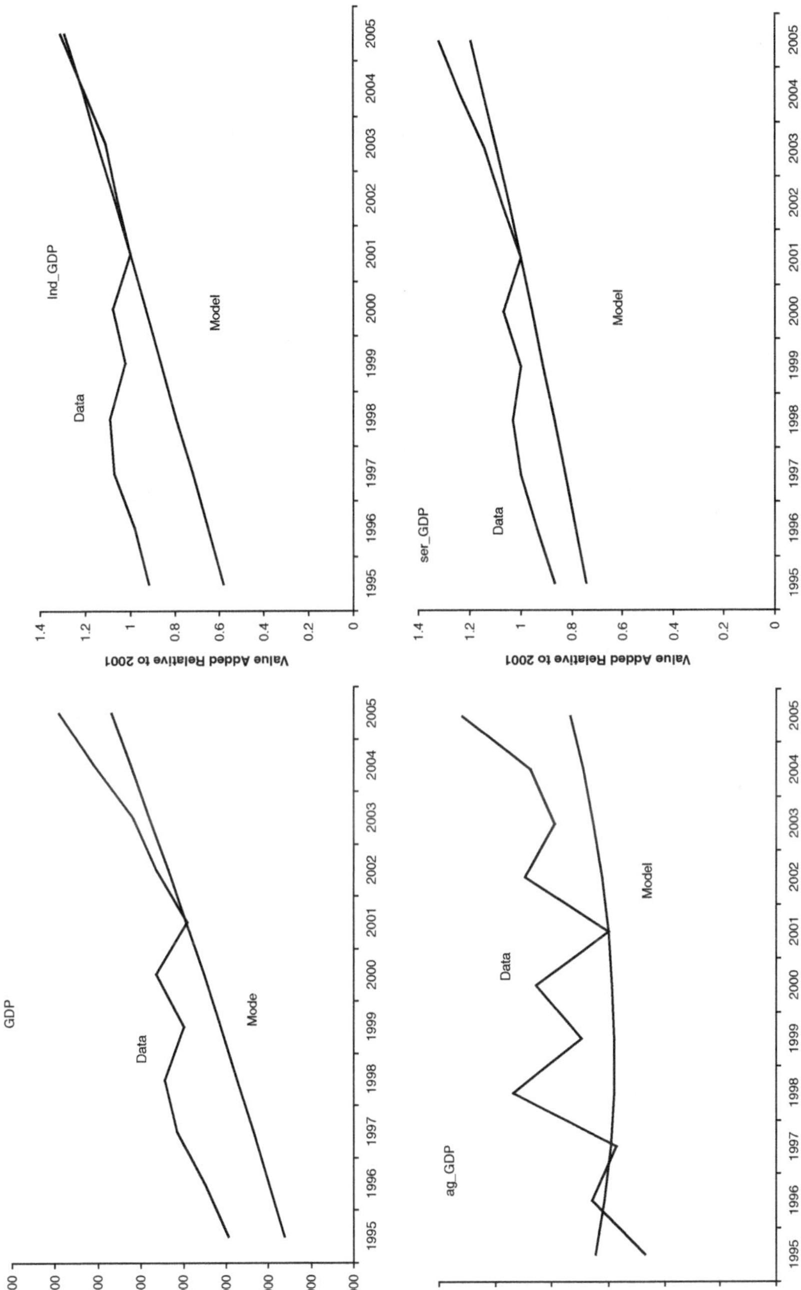

Figure 6.1 Confronting model forecasts to the data

forecast, and higher for industry due to the relatively large difference between the normalized values over the 1995–2001 period. Theil's U statistic is unbounded from above with smaller values indicating a closer fit to the data.[9] This measure also tends to show the predicted values for agriculture to be lower than is the case for the other sectors.

Table 6.2 Measures of the model's forecast accuracy, 1995–2005

Measure	Economy GDP	Agriculture GDP	Industry GDP	Service GDP
Correlation Coefficient	0.900	0.457	0.811	0.928
Concordance Correlation Coefficient	0.610	0.133	0.429	0.666
Theil's U Statistic	0.167	0.052	0.193	0.106
Mean Absolute Error (%)	10.737	4.209	15.070	9.633

6.2.3 Empirical results

Adding government and composite capital to the empirical model of Chapter 5 leads to a slightly decreased rate of growth relative to the model with only intermediate inputs.[10] The model predicts a rate of growth in GDP per worker equal to 2.91 percent in year 2002. This rate is 0.12 percentage points less than the intermediate inputs model of the preceding chapter. The slower rate of growth causes total factor payments per worker to double in about 30 years as opposed to 28 years. By 2012, the slightly slower growing economy reaches its half-way point of steady-state growth in GDP per worker of 2.4 percent per annum. By

[9] The unweighted Theil U statistic is given by $\sqrt{\sum_t \left(Y_t - \hat{Y}_t\right)^2 / \sum_t Y_t^2}$ where $t = 1995, \cdots, 2005$.

[10] The economy and sectoral value added results presented in this chapter include indirect production taxes which comprise over 58 percent of government tax revenues. Some caution should thus be exercised in making direct comparisons of the value added reported in this chapter with previous chapters. Factor income comparisons are appropriate.

year 2050 the economy is within 2.8 percent of its long-run rate of growth, at which point income per worker has tripled. Table 6.3 shows the ratio of the economy's capital stock to total factor income increases from its initial level of 4 in 2001, to 4.52 in 2031. Capital per worker increases by a factor of 1.4 between 2001 and 2011, and by a factor of 2.3 by the year 2031. These values are slightly smaller than the values predicted by the previous model.

The pattern of factor payments is similar to the preceding example, with land rental income declining in the early periods and then increasing by 2016. Relative factor payments depart from the previous chapter in that payments to labor are slightly higher here, while capital earnings and land rental payments are slightly lower. Expenditure per worker is lower than reported in all of the previous examples because government consumption is treated separately from that of the household. In spite of the greater structural detail of this extended model, the general pattern of factor income and expenditure over time is unchanged from that of the basic model of Chapter 4.

The evolution of sectoral value shares in GDP, and resource shares in each sector are reported in Table 6.4. The model reproduces initial sector value shares in GDP that correspond closely to the values reported by the data (see Chapter 8). As the internal terms of trade evolve and capital deepening occurs, the share of the industrial sector in GDP increases from 20.6 percent in 2001 to 26 percent by 2031. Agriculture's share in GDP declines over this period, while the service sector's share of GDP increases modestly from 63.6 to 64.7 percent. The relative rise in the value share of the industrial sector in GDP leads to a trade reversal in about 2014, as the country begins to export the industrial good. In spite of the Cobb-Douglas specification for composite capital, $F(y_{mk}(t), y_{ak}(t), y_{sk}(t))$, the substitution away from service output in capital is small due to the relatively modest 1.3 percent rise in the relative price of the service good over the 2001–2031 period. The ratio of the quantity of the service to capital good in composite capital falls from 1.065 initially to 1.051 in 2031.

Table 6.3 Factor income and expenditure in millions of 2001 Turkish Lira

Year	GDP per worker	Capital per worker	Wage income per worker	Capital earnings per worker	Land rental income per worker	Expenditure per worker	Saving to GDP
2001	7442	25896	3342	3105	82	3558	0.38
2006	8635	31168	3831	3606	80	4138	0.37
2011	9861	36550	4341	4119	80	4738	0.36
2016	11141	42103	4881	4656	83	5367	0.36
2021	12494	47892	5455	5222	87	6031	0.36
2026	13936	53981	6070	5825	92	6741	0.35
2031	15485	60450	6734	6474	99	7500	0.35

Source: Model results

In contrast to the results of the numerical example of Chapter 5, along the transition path the industrial and the agricultural sector's rates of growth in gross output are marginally higher in this example, while the service sector's rate of output growth is modestly lower. The capital stock per worker reported in Table 6.3 is marginally lower than the values reported in Table 5.3 while in contrast GDP per worker is slightly higher here than the values reported in Table 5.3. Identifying precisely the reasons for these somewhat minor differences requires a more exhaustive analysis than we undertake here. Note also that the sectoral value shares in GDP predicted here correspond more closely to the data reported in Chapter 8 than do the corresponding shares predicted by the model of the preceding chapter. However, it can be shown that by choosing other "reasonable" levels of initial capital stock and a rate of growth x in labor services – parameters that are problematic to estimate from the data – the empirical models of both Chapters 4 and 5 can provide a better fit to the data.

Changes in sectoral factor shares correspond to changes in sectoral GDP shares, but these changes tend to obscure the substitution of capital for labor as wages rise and capital rental rates fall. The ratio of capital to labor employed in agriculture increases from a factor of 1.2 between 2001 and 2006 to 2.3 over the 2001–2031 period. The share of labor employed in agriculture falls, as does the total quantity of workers employed in the sector. The net effect of these adjustments is an increase in land rent per agricultural worker throughout transition growth, even though the land rental income per economy wide worker reported in Table 6.3 shows a non-monotonic pattern. Land income per farm worker rises by a factor of 1.15 over the period 2001–2006, and by a factor of 2.0 over the 2001–2031 period. These values are almost identical to the results in Chapter 4.

Tables 6.5, 6.6, and 6.7 illustrate the effect of changes in value-added prices and capital deepening, on the growth in sectoral gross output. These growth accounting exercises are performed using the empirical counterpart of the supply functions (6.19), (6.27) and (6.28). Tables 6.5 and 6.7 show the Rybczynski-like

Table 6.4 Sector value shares in GDP and sector factor shares in total factors

	Sector Share in GDP			Labor Share in			Capital Share in		
Year	Industry	Agriculture	Service	Industry	Agriculture	Service	Industry	Agriculture	Service
2001	0.206	0.158	0.636	0.175	0.167	0.658	0.244	0.126	0.630
2006	0.227	0.133	0.640	0.194	0.142	0.665	0.266	0.106	0.628
2011	0.239	0.118	0.642	0.205	0.126	0.669	0.280	0.093	0.627
2016	0.248	0.108	0.644	0.213	0.115	0.672	0.289	0.085	0.626
2021	0.254	0.101	0.645	0.218	0.108	0.674	0.295	0.079	0.626
2026	0.257	0.096	0.646	0.222	0.103	0.675	0.299	0.075	0.626
2031	0.260	0.093	0.647	0.224	0.099	0.677	0.302	0.073	0.626

Source: Model results

effects of capital stock and effective labor growth on industrial and service sector gross output. These Rybczynski-like effects can be expressed roughly as follows: Capital deepening has the effect of increasing the productivity of labor and, all else constant, lowering the cost of production to industrial sector firms. Industrial firms respond by increasing their demand for labor, which in turn places upward pressure on wages. As the service sector is labor intensive, the increase in wages dominate the fall in service sector capital rental costs, and hence, induces an increase in service sector production costs.

Three factors account for the growth in service sector demand, (i) the sector accounts for about half of composite capital; (ii) the sector's output constitutes a relatively large share of intermediate demand; (iii) the increase in household and government service demand that accompanies increases in real income over time. Together, the growth in total service sector demand induces an increase in the service sector price, which enables service sector firms to profitably employ additional resources for the service good market to clear.

Table 6.5 Growth in service gross output and factor contributions

		Contributions to Growth		
	Growth in Gross output	Value added price	Capital stock	Effective labor
2001	0.0448	0.0861	−0.1925	0.1512
2006	0.0414	0.0608	−0.1709	0.1515
2011	0.0392	0.0437	−0.1562	0.1516
2016	0.0377	0.0317	−0.1458	0.1517
2021	0.0366	0.0232	−0.1385	0.1518
2026	0.0358	0.0169	−0.1330	0.1518
2031	0.0352	0.0126	−0.1292	0.1519

Source: Model results

The home-good price increase leads to a decline in the internal terms of trade for the industrial and agricultural sectors. It also triggers a modest increase in the unit price of capital, accounting for about one-half of the 1.3 percent rise in the service good price over the 2001–2031 period. The decline in the internal terms of trade dampens the traded goods sector's demand for resources. This dampening effect is greater for the industrial sector than for agriculture, as service sector output constitutes a larger share of gross industrial production than it does in gross agricultural production.

The decline in the capital rental rate has the effect of lowering agriculture's cost of production. However, since capital is second to labor in agriculture's total production costs, the growth in wage costs dominates the fall in capital costs, even though, as noted above, the sector substitutes capital for labor. As the rate of increase in wages, and the rate of decrease in capital costs dissipate, agricultural output grows to eventually converge to the same rate of long-run growth of the other two sectors. Of course, this outcome would not obtain without the 3.36 percent rate of growth in land productivity.

Table 6.6 Growth in industrial gross output and factor contributions

		Contributions to Growth		
Year	Growth in Gross output	Value added price	Capital stock	Effective labor
2001	0.0707	−0.2440	0.7217	−0.4071
2006	0.0558	−0.1595	0.5856	−0.3703
2011	0.0480	−0.1098	0.5085	−0.3507
2016	0.0434	−0.0776	0.4600	−0.3390
2021	0.0404	−0.0558	0.4278	−0.3316
2026	0.0384	−0.0403	0.4054	−0.3266
2031	0.0371	−0.0296	0.3900	−0.3233

Source: Model results

6.3 Trade reform

The structural detail of the model here allows for a number of policy experiments. We examine the impact of a once-and-for-all 20 percent decrease of the ad valorem tariff rate on industrial good imports, and leave unchanged indirect production taxes.

Table 6.7 Growth in agricultural gross output and factor contributions

		Contributions to Growth			
Year	Growth in Gross output	Value added price	Wage effect	Interest rate effect	Technical change
2001	0.0050	−0.0028	−0.0658	0.0400	0.0336
2006	0.0129	−0.0021	−0.0477	0.0290	0.0336
2011	0.0185	−0.0015	−0.0348	0.0212	0.0336
2016	0.0225	−0.0011	−0.0255	0.0155	0.0336
2021	0.0254	−0.0008	−0.0188	0.0115	0.0336
2026	0.0276	−0.0006	−0.0138	0.0084	0.0336
2031	0.0292	−0.0005	−0.0103	0.0063	0.0336

Source: Model results

Decreasing the tariff has a small effect on government receipts, as tariff revenues only account for 4.3 percent of total receipts in the base period data, see Table 6.1. Still, as suggested by the government transfer Equation (6.24), the likely shortfall in government revenue that results from decreased tariff revenue is offset by increased transfers from households.

We alert the reader to two issues associated with the tariff experiment. First, since indirect production taxes induce other distortions in the economy, a decrease in the level of protection afforded the industrial sector will not necessarily increase welfare or aggregate GDP. Second, by decreasing the tariff rate, we change value-added and border prices relative to the prices that prevail in the pre-reform scenario (the base solution). To address the price regime problem, we use Equation (6.1) and compare the numerical value of the discounted present value of utility obtained from the base solution with the corresponding value obtained from the simulation. Our interest is in the rank order of these values as opposed to the magnitude of their numerical values. The value obtained from the base solution is smaller than the value obtained from the simulation. We thus conclude that a reduction in the tariff rate by 20 percent is welfare improving: Of course, a different discount rate could potentially yield a different ranking of scenarios.

The smaller discounted present value of utility accompanying the base solution, however, does not imply felicity obtained from the simulation is greater than felicity obtained from the base solution for all t. Indeed, between 2001 and 2026, the felicity values from the base solution are smaller than the corresponding values from the simulation. Beyond 2026, the base solution felicity values do not sufficiently exceed felicity values from the simulation to overcome the time rate of discount ρ used to calculate the discounted present value of utility.

Real expenditure per worker is readily calculated using the expenditure function (6.3). From this equation we obtain the price index $\mathcal{E}(p_a, p_m, p_s(t))$ of aggregate consumption q. If we divide nominal expenditures per worker by the expenditure function component $\mathcal{E}(p_a, p_m, p_s(t))$ we obtain real expenditure measured as the quantity index $q(t)$. Performing this calculation for both the base solution and the simulation, and then dividing the latter by the former leads to values greater than unity through about the year 2020 (see Table 6.8). This result indicates that households consume a larger amount of aggregate consumption when the tariff rate is reduced compared to the base solution.

While we conclude lowering the industrial good tariff rate by 20 percent is welfare improving, it is not the case that reducing the tariff leads to higher GDP per worker, measured at constant prices. The first column of Table 6.8 is calculated as follows. The quantities of final goods obtained from the simulation are evaluated at constant initial period prices. The same constant prices are used to calculate the value of final goods obtained from the base solution. The value obtained from the simulation is then divided by the corresponding value from the base solution. The result, reported in the first column of Table 6.8, shows the value of the base solution GDP calculated in this manner exceeds GDP obtained from the simulation. The time required for the GDP values to double on a per worker bases is roughly the same – 28.4 years for the base solution, and 29.8 years for the simulation.

Factor income is calculated differently. For the base solution we divide each source of income by this solution's price index $\mathcal{E}(p_a, p_m, p_s(t))$. The same calculation is performed with the simulation, using the price index obtained from the simulation. Then, the resulting values calculated for the simulation are divided by the values from the base solution. The results are also reported in Table 6.8.

The results show that decreasing industrial sector protection cause real wages to rise over the 2001–2006 period, earnings from capital fall, while land rental income increases. The increase in real land rental income associated with the decreased protection is over twice the level of land rental income realized in the base solution. The decline in capital earnings is linked to the decline in the real value of the capital asset, measured as $p^k(t)\, k(t) / \mathcal{E}(p_a, p_m, p_s(t))$. The ratio of this value obtained from the simulation divided by the base solution value falls from 0.96 in 2006 to 0.91 by 2031. The ratio of the simulation to base solution of the normalized value of land, Equation (6.11), is 2.37 in 2006 and remains virtually unchanged over the 2006–2031 period. Nevertheless, real GDP as the sum of factor earnings is lower in the post-reform economy.

If real income from the reduction in tariffs is less than real income obtained from the base solution, how can the representative household be better off? The answer is that a reduction in the rate of protection is a terms of trade change that favors the less capital intensive sectors of the economy. Households save a smaller share of their income and spend a larger share of income on final goods. The ratio of saving to GDP in the simulation is smaller than the same ratio in the base solution (see the last column of Table 6.8).

Close examination of the reduced form factor rental rate equations (6.25) and (6.26), reveal that a change in the tariff rate affects the initial level of the value-added prices p_{vm} and p_{vs}, as well as the value of p_s^{ss}, the root satisfying the Euler condition (6.34) in the steady-state. The value of p_s^{ss} from the base solution exceeds the value from the simulation by 5 percent. The steady-state level of capital stock from the base solution

Table 6.8 Effects of a reduction in the tariff rate of industrial good imports

	Ratio of simulation values to base model values					
Year	GDP	Wage income	Capital earnings	Land rental income	Expenditure	Saving to GDP
2001	0.9772	1.0232	0.9326	2.0845	1.0378	0.9204
2006	0.9658	1.0078	0.9232	2.1854	1.0193	0.9308
2011	0.9578	0.9970	0.9169	2.2627	1.0090	0.9335
2016	0.9522	0.9895	0.9128	2.3188	1.005	0.9399
2021	0.9486	0.9844	0.9104	2.3577	0.9929	0.9448
2026	0.9463	0.9811	0.9091	2.3840	0.9887	0.9480
2031	0.9448	0.9788	0.9084	2.4022	0.9860	0.9499

See text for the methods used to calculate these values

exceeds the stock level from the simulation by almost 8 percent. Applying the relative rate of protection formula (6.14), the industrial sector experiences an initial negative terms of trade effect of 0.9 percent while agriculture experiences a positive terms of trade effect of 6.3 percent. A reduction in the tariff rate thus has the direct effect of lowering the price received by firms in the industrial sector, and the indirect effect of lowering the cost of industrial goods employed as intermediate inputs of production in the other sectors. Another secondary effect is the lower the price of the home-good, the lower the cost of the service good employed as an intermediate factor of production in each sector.

The effect on the structure of the economy from reduced industrial sector protection is shown in Table 6.9. Decreased protection causes industrial sector firms to release resources to the agricultural and service sectors of the economy. The model predicts, in the initial period, industry's share in final GDP falls from 20.6 percent in the base solution to 6.2 percent in the simulation, while in 2031, industry's share falls from 26 percent in the base solution to 14 percent in the simulation (see Table 6.4). These large changes, at least in the initial period, likely do not reflect a real economy's response to the change in incentives brought about by the change in prices. The model reallocates labor among sectors of the economy as though there are no labor market frictions, and reallocates capital stocks across sectors in a costless fashion. A shock to the model of a sufficiently small magnitude that only slows the growth in sectoral capital stock is likely to provide more reliable predictions of sectoral adjustment than when the shock is of a magnitude to induce a sector to reduce its stock of capital.[11]

The incentive for agricultural firms to increase production relative to the base solution is due to the initial 6.3 percent improvement in its terms of trade, and the lower nominal wage and lower capital rental rate. Thus, agricultural firms experience an improvement in terms of trade and a decline in production costs. As a result, the economy increases agricultural exports, and increases industrial good imports, which causes the share of the value of exports plus imports to real GDP to exceed the share of GDP traded in the base solution.

However, capital deepening has the same sectoral effects on the evolution of production as those discussed for the case of the base solution. The Rybczynski-like forces associated with capital deepening on the growth in sectoral gross output are the same as the base model, although the magnitude of these effects tend to vary from the base model. The effects of capital deepening increase the productivity of labor in the industrial sector which causes the sector to demand more labor. Roughly, agricultural and service sector firms experience an increase in the cost of production because the decline in the capital rental rate is not sufficient to compensate for the increase in cost due to rising wages. The growth in final and intermediate demand for the service good requires the price of the service good to rise, which in turn provides the incentive for service sector firms to profitably compete for the labor and capital to increase supply

[11] See Barro and Sala-i-Martin (2004), Chapter 3 for discussion of capital adjustment costs and Romer (2006) Chapter 8 for a discussion of kinked and fixed adjustment costs. An easy way to slow the growth in capital stock in our model with composite capital is to require firms to employ some of the economy's resources in forming the composite capital as opposed to the treatment here, where the cost of the composite only depends on the prices of final goods.

Table 6.9 Effects of a reduction in the tariff rate of industrial good imports

	Sector Share in GDP			Labor Share in			Capital Share in		
Year	Industry	Agriculture	Service	Industry	Agriculture	Service	Industry	Agriculture	Service
2001	0.062	0.331	0.607	0.051	0.339	0.609	0.078	0.282	0.640
2006	0.090	0.298	0.613	0.075	0.307	0.619	0.112	0.250	0.638
2011	0.108	0.275	0.616	0.090	0.285	0.625	0.134	0.230	0.637
2016	0.121	0.266	0.619	0.101	0.270	0.629	0.150	0.215	0.636
2021	0.130	0.249	0.621	0.109	0.258	0.632	0.160	0.205	0.635
2026	0.137	0.240	0.623	0.115	0.250	0.635	0.168	0.198	0.635
2031	0.142	0.234	0.624	0.120	0.244	0.637	0.174	0.192	0.634

Source: model results

of the service good. Over the years 2001 through 2031, the rise in the service good price is only 0.8 percent, which causes the deterioration in agriculture's domestic terms of trade to be less than the deterioration experienced in the base solution over the same period. Nevertheless, labor departs agriculture while capital deepening per worker in the sector increases. As in the base solution, growth in agricultural output per worker converges from below to the long-run balanced growth rate of 1.9 percent per annum. The effect on trade is to decrease the share of GDP traded to 31 percent by 2031.

6.4 Conclusion

This chapter took up the task of combining into one model each of the extensions to the three-sector model presented in Chapter 5. We also extended the no-arbitrage condition of the preceding chapter to derive the price of land. Combining each of the 5 extensions was shown to increase the complexity of the model, while making it more suited to address a broader array of policy analyses. Disaggregating the economy to account for a more narrow definition of sectors only creates the additional problem of sector closure in transition growth, but otherwise does not raise barriers beyond those encountered in this chapter.

The empirical exercise was divided into a validation and an empirical results section. We suggested that validating the model's predictions by confronting them to data is essential if policy makers and others are to have confidence in the predictions of the modeled economy. We suggest validation exercises can help make the applied dynamic general equilibrium analyses of the type considered here, a scientific discipline in which there are well-defined puzzles with clear successes and failures of competing theories. At the same time we suggest that models of complex phenomena (like real economies) should be viewed as a metaphor, and their value is determined on the basis of how useful they are as a tool to accomplish an end.

We discussed a number of issues encountered when fitting the model to data, and empirical measures for evaluating the results. The issues included the problem of fitting the model to an extreme point of a country's business cycle, the problems that arise when time series data do not correspond to the definition of model variables, and how these concerns might be addressed. While the validation measures we presented are well known in the literature, we briefly mentioned others and noted that much work needs to be done in this regard.

The numerical results indicate the model provides a relatively good fit to the data of a country that has experienced rather substantial business cycles over the 1995–2005 period. The predicted values of many variables however were found to be close to the predicted values of the less complex models in the preceding two chapters – this is especially the case for the rate of growth of GDP and the time it takes to double income per worker. The more detailed model of this chapter, however, was more accurate in predicting sector shares in GDP. We noted that the empirical results of the other models could be improved by changing the values of parameters prone to being mismeasured, e.g., capital stock and the initial rate of return to capital. Of course, changing parameters raises the question of consistency across models.

The empirical exercise is concluded by conducting a simulation which reduced the tariff rate protecting industrial sector firms by 20 percent, and contrasting the result with base solution results. We noted a potential weakness of the model in that industrial sector firms reduced output initially to a greater extent than might be expected in a real economy. This weakness was attributed to the costless transfer of resources between sectors, and in particular, the reallocation in early periods of capital stock from the industrial sector to other sectors of the economy. Remedies to this problem were suggested. In spite of the reallocation issue, the basic forces discussed in the base model prevailed in the simulation with agriculture's share in GDP falling, and labor and capital share of total labor and capital, respectively, falling in agriculture and rising in the other two sectors. The change in the terms of trade between the industrial and agricultural sector due to the reduction in the tariff rate caused an initial decline in the price of the service good relative to the price of traded goods. The slower rate of increase in the service good price over time relative to the base model, helped to increased the share of the country's GDP traded internationally.

The next chapter draws upon the three-sector model of Chapter 4 and relaxes the assumption that the rest of the world is in a steady-state equilibrium. We consider a two-country-world, and present two models. The first model allows for international capital flows, so the rate of return to capital in the home country and the rest of the world is equalized. The second model disallows capital flows between countries, which results in a model with two state variables. This extension lays the foundation for developing multi-country models with structural features presented in this and the preceding chapter.

7

A Three-Sector – Two-Country World

Chapters 4, 5, and 6 develop models of a small open economy facing exogenous and constant traded good prices. Implicit in these models is the assumption that the rest of the world is in a steady-state equilibrium. Moreover, these models restrict ownership of capital stock to domestic residents, and preclude them from holding other country assets. Consequently, own saving is the sole source of domestic investment and the value of a country's exports equal the value of its imports. We now relax these two assumptions and extend the basic model of Chapter 4 to a two-country world.

Two separate models are constructed. The first models a two-country world with perfect capital mobility. Capital mobility means that domestic and foreign claims on capital are perfect substitutes as stores of value so that a single world interest rate $r(t)$ at each t prevails. An implication is that domestic assets per worker need not equal a country's capital stock per worker and capital flows between countries can cause one country to incur a trade surplus and another a trade deficit that prevails in the long-run. The other country can be treated as the rest of the world or as a conglomerate of a country's trading partners. Agents in both countries are assumed to behave as though they take international prices and rates of return to capital as given, hence omitting concern with strategic behavior.

For some empirical applications, a model with capital mobility in a two-country world may poorly depict the actual environment due to various policy interventions, institutional constraints or other capital market barriers. We thus develop a second model with the same production and consumption structure as the first, but agents of one country are not allowed to

T.L. Roe et al., *Multisector Growth Models*, DOI 10.1007/978-0-387-77358-2_7,

own capital stock of the another country. In this second model, the rate of return to countries' capital can differ in transition to long-run equilibrium. In the long-run however, our assumptions regarding household preferences cause the returns to capital to converge to a common steady-state value.

Linking countries through trade in final goods and capital speeds up the transition growth of the country whose initial capital-labor ratio is relatively low, while the reward to the net-lending country is the remuneration to savings invested in the former. An international capital market also has profound effects on the co-movement of the price of home-goods and wage payments among countries. The basic models presented here serve as a point of departure for analyzing of the effects of policies such as thoughs that distort border prices, place a tax on capital earnings, or constraints on foreign liabilities as in Barro et al. (1995), and other interventions in one country that have both direct and indirect effects on growth of the other country.

The presence of a home-good in each country allows for other considerations. For example, the theory of international trade typically assumes that competition induces countries to employ the same technology in the production of traded goods so that comparative advantage arises from the relative differences in their factor endowments. International competition is less likely to induce countries to employ the same technology in the production of non-traded goods. An argument can thus be made for trade liberalization among countries in the production of mostly service type goods, and policy to encourage the more efficient production of goods that by their nature are costly to trade internationally. Efficiency gains in the production of these goods can release resources for the production of traded goods. The models of the type presented here can be used to provide insights into issues of this nature.

The framework presented here touches on a number of other issues in the literature. The closed economy Ramsey model of Chapter 3 predicts that the level of capital stock to which a country converges to is independent of its initial capital stock. Hence, countries that differ only in their initial levels of capital stock should converge to the same steady-state and, consequently, share common levels of income in the long-run. However, Atkeson and Kehoe (2000) and Gaitán and Roe (2007) show that this is not necessarily the case. Atkeson and Kehoe consider a country that has the same technologies and inter-temporal preferences as the rest of the world, but begins the development process with a capital-labor ratio lower than the rest of the world. They show that long-run equilibrium is not independent of the initial conditions. Gaitan and Roe show for otherwise identical countries, the country with an initial lower capital-labor ratio will converge to long-run income level lower than would prevail under autarchy, while the other country will converge to an income level higher than under autarchy.

We develop both models in a manner analogous to that found in the previous chapters. To keep the notation manageable, we make a number of simplifying assumptions. Most importantly, countries are assumed to employ the same technology in the production of all goods, and households hold identical preferences. Thus, countries only differ in their endowments of labor, capital and land. These assumptions are relaxed in the empirical exercise which is used to illustrate some key properties of the model. One major technical difference between the first and second model is the former model has a single state variable describing the evolution of the world's stock of capital, while the second model involves two state variables: one for each country's capital stock. We show a convenient way to deal with this two state variable problem.

7.1 A two-country world with capital mobility

The world is composed of two countries, where the economic environment of each country is virtually identical to that of the three-sector model presented in Chapter 4. Each country employs the same technologies and households hold identical preferences. This means, for example, if the production technology for sector j in the home country is given by $Y_j =$

$\Psi_j (L_j)^\alpha (K_j)^{1-\alpha}$, then the production technology for sector j in the other country is also given by $Y_j^o = \Psi_j \left(L_j^o\right)^\alpha \left(K_j^o\right)^{1-\alpha}$.

To distinguish between the "home" and "other" country, we assign an "o" superscript to the other country variables and leave the home country variables un-marked. The home country initial conditions are labor $L(0)$, capital $K(0)$ and land H while for the other country they are $L^o(0)$, $K^o(0)$ and H^o. We assume no technical change, and the labor force grows at the same rate n per annum in both countries. We further simplify the notation by only considering the case where the inter-temporal elasticity of substitution $1/\theta$ is unity. Assuming this degree of similarity is unnecessary in empirical application, but it greatly simplifies our notation.

Agents in each country take world market prices as given, and capital is mobile between countries. We assume domestic and foreign claims on capital are perfect substitutes as stores of value so that a single world interest rate $r(t)$ prevails at each t. An implication of the capital mobility assumption is, at each instant in time, the marginal value product of capital is equated across countries. Each country produces an agricultural good, a manufacturing good and a home-good. The agricultural and home-good enter final consumption only, with any excess supply or demand of the agricultural good exported or imported at a world market clearing price p_a. The manufacturing good enters final consumption, and contributes to the country's stock of capital. Any excess supply or demand of the manufacturing good is exported to or imported from the other country at the world numeraire price of unity.

7.1.1 Households and firms

The capital stock held by households in models presented in Chapters 4, 5, and 6 equal the stock of capital employed by firms. Here, the stock of capital held by home country households can depart from the total stock of capital employed by home country firms if households in the other country hold claims to some of the capital in the home country. Let K_A denote the amount of capital stock held by home country households. This stock of capital equals the difference between domestic capital stock K, less the net claims D on K held by households in the other country

$$K_A = K - D$$

If D is negative, then home country residents hold claims on other country capital K^o. The home country's flow budget constraint is

$$\dot{K} - \dot{D} = wL + r(K - D) + \Pi H - E \tag{7.1}$$

which remains unchanged from Chapter 4 except for the presence of $\dot{D}$ and D. The definition of the remaining terms are unchanged from previous chapters.

For the foreign country we have

$$\dot{K}^o + \dot{D} = w^o L^o + r(K^o + D) + \Pi^o H^o - E^o \tag{7.2}$$

where in this case

$$K_A^o = K^o + D$$

Hence, the world's total stock of capital is

$$K^w = K_A + K_A^o$$

Assuming that domestic and foreign claims on capital are perfect substitutes as stores of value – so that a single world interest rate $r(t)$ prevails at each t – leads not only to the no-arbitrage condition between capital and land within a country, as shown in Chapter 3, but also to a no arbitrage condition for land between countries.

Note that, although D is the same value in both countries, on a per worker basis, the net claim to labor ratio, D/L, of the home country is not equal to the corresponding value D/L^o of the other country. Thus, variables must be normalized by the total number of world workers. A bar "-" is used to indicate a variable is expressed in per world worker terms.

Since the structure of household preferences are identical in the two countries, the results from forming the present-value Hamiltonian leads to the familiar Euler condition

$$\frac{\dot{\epsilon}}{\epsilon} = \frac{\dot{\epsilon}^o}{\epsilon^o} = (r - \rho)$$

for the case of unitary inter-temporal elasticity of substitution. This condition is the same for both countries so their rate of change in expenditures ϵ per country worker are equal. Moreover, since the labor force in both countries grows at the same rate n, which we assume is in the same proportion to the growth in population, the rate of change in expenditure per representative world worker can also be expressed as

$$\frac{\dot{\bar{\epsilon}}}{\bar{\epsilon}} = (r - \rho) \tag{7.3}$$

where the "–" notation denotes per world worker.

Technologies remain unchanged from previous chapters, except for the need to express these relationships in per world worker terms. Consequently, cost functions for home country firms are

$$C^j\left(w, r^k\right) \bar{y}_j \quad j = m, s,$$

where $\bar{y}_j = Y_j / (L + L^o)$. Total value-added by agriculture's sector specific resource is given by $\pi(p_a, w, r)\, \mathcal{B}(t)\, H$, where the sustainability condition $\mathcal{B}(t) = e^{nt}$ is assumed. In per world worker terms, value-added is

$$\boldsymbol{\pi}^a\left(p_a, w, r^k\right) \bar{H}$$

where the sector specific endowment is now stated as $\bar{H} = \ell H$, and $\ell = L/(L + L^o)$ is the share of own country workers in total world workers.

Similarly, we have for the other country firms

$$C^j\left(w^o, r^k\right) \bar{y}^o_j \;\; j = m, s \text{ and } \boldsymbol{\pi}^a\left(p_a, w^o, r^k\right) \bar{H}^o$$

where $\bar{H}^o = (1 - \ell)\, H^o$.

Before characterizing the equilibrium, it is useful to review some basic identities implied by the model.

7.1.2 Basic identities

The net claims of households in one country on the capital stock of another country causes the net value of trade in the two countries to differ by the net remuneration households receive on these claims. To show this linkage, express the own country budget constraint (7.1) in terms of world workers

$$\dot{\bar{k}} - \dot{d} = w\ell + \bar{k}\left(r^k - n - \delta\right) - d\left(r^k - n - \delta\right) + \pi\bar{H} - \bar{\epsilon}$$

Here, $\bar{k} = K/\left(L + L^o\right)$ and the remaining variables are defined similarly. We omit a "-" on d because it is the same value in both countries. Competition among firms implies the value of output is exhausted by payments to factors

$$\bar{y}_m + p_a\bar{y}_a + p_s\bar{y}_s = w\ell + r^k\bar{k} + \pi\bar{H}$$

Replace factor payments in the budget constraint by the value of output, and reorganize terms to obtain the home country's balance of payments condition

$$d\left(r^k - n - \delta\right) - \dot{d} = \bar{y}_m - \bar{q}_m - \dot{\bar{k}} - \bar{k}\left(n + \delta\right) + p_a\left(\bar{y}_a - \bar{q}_a\right) \quad (7.4)$$

where expenditure is

$$\bar{\epsilon} = \bar{q}_m + p_a\bar{q}_a + p_s\bar{q}_s$$

Clearing of the home-good market causes the terms $p_s\bar{y}_s$ and $p_s\bar{q}_s$ to cancel.

The right hand side of expression (7.4) is the country's trade balance, which equals capital payments the country pays or receives, adjusted for the change in net claims. If d is positive and $\dot{d} = 0$, as we might expect in long-run equilibrium, a trade surplus is required to remunerate foreigners' claims d on domestic capital stock. In this case, the country incurs a trade surplus forever.

Expressing the other country's budget constraint (7.2) in terms of world workers gives

$$\dot{\bar{k}}^o + \dot{d} = w^o\left(1 - \ell\right) + \bar{k}^o\left(r^k - n - \delta\right) + d\left(r^k - n - \delta\right) + \pi^o\bar{H}^o - \bar{\epsilon}^o$$

where $(1-\ell)$ is the share of other country workers in total world workers. Proceeding as above we obtain

$$\dot{d}-d\left(r^k-n-\delta\right)=\bar{y}_m^o-\bar{q}_m^o-\dot{\bar{k}}^o-\bar{k}^o\left(n+\delta\right)+p_a\left(\bar{y}_a^o-\bar{q}_a^o\right) \quad (7.5)$$

In the long-run, where $\dot{d}=0$, if d is positive the other country incurs a trade deficit that can persist forever. In this case, the country's gross national product is greater than its gross domestic product and households in this country can consume in excess of the earnings from domestic resources alone.[1]

Summing (7.4) and (7.5) causes the debt terms to cancel and leads to the world condition that the value of excess supply equals the value of excess demand. This result also implies the world's flow budget constraint is equal to

$$\dot{\bar{k}}^w=w\ell+w^o\left(1-\ell\right)+\bar{k}^w\left(r^k-n-\delta\right)+\pi\bar{H}+\pi^o\bar{H}^o-\bar{\epsilon}^w \quad (7.6)$$

where world capital stock and expenditure per world worker are

$$\begin{aligned}\bar{k}^w &= \bar{k}+\bar{k}^o\\ \bar{\epsilon}^w &= \bar{\epsilon}+\bar{\epsilon}^o\end{aligned}$$

Equation (7.6) is one of three differential equations used to solve for inter-temporal equilibrium values.

7.1.3 *Equilibrium*

For a given sequence $\left\{\bar{k}^w\left(t\right),\bar{\epsilon}\left(t\right),\bar{\epsilon}^o\left(t\right)\right\}_{t\epsilon[0,\infty)}$ of capital and expenditure pairs and given initial resource endowments

$$\left\{K\left(0\right),L\left(0\right),H,K^o\left(0\right),L^o\left(0\right),H^o\right\}$$

an intra-temporal equilibrium consists of the ten-tuple sequence of positive values

$$\{w\left(t\right),w^o\left(t\right),r^k\left(t\right),p_s\left(t\right),p_s^o\left(t\right),p_a\left(t\right),$$
$$\bar{y}_m\left(t\right),\bar{y}_m^o\left(t\right),\bar{y}_s\left(t\right),\bar{y}_s^o\left(t\right)\}_{t\epsilon[0,\infty)}$$

---

[1] Here, GNP = GDP less net factor payments to the rest of the world. Since the only factor that is mobile across countries is capital, the difference between GNP and GDP in the long-run is the net capital earning to the other country.

satisfying the ten equations (7.7), (7.8), (7.9), (7.10), (7.11), (7.12), (7.13), and (7.14) for each t

- zero profits in manufacturing and the home-good production,

$$\begin{aligned} C^m\left(w, r^k\right) &= 1, \quad C^m\left(w^o, r^k\right) = 1 && (7.7) \\ C^s\left(w, r^k\right) &= p_s, \quad C^s\left(w^o, r^k\right) = p_s^o && (7.8) \end{aligned}$$

- labor market clearing in home and other country, respectivcly

$$\begin{aligned} \sum_{j=m,s} C_w^j\left(w, r^k\right)\bar{y}_j - \boldsymbol{\pi}_w^a\left(p_a, w, r^k\right)\bar{H} &= \ell && (7.9) \\ \sum_{j=m,s} C_{w^o}^j\left(w^o, r^k\right)\bar{y}_j^o - \boldsymbol{\pi}_{w^o}^a\left(p_a, w^o, r^k\right)\bar{H}^o &= (1-\ell) && (7.10) \end{aligned}$$

- world capital market clearing[2]

$$\sum_{j=m,s} C_{r^k}^j\left(w, r^k\right)\bar{y}_j + \sum_{j=m,s} C_{r^k}^j\left(w^o, r^k\right)\bar{y}_j^o -$$

$$\boldsymbol{\pi}_{r^k}^a\left(p_a, w, r^k\right)\bar{H} - \boldsymbol{\pi}_{r^k}^a\left(p_a, w^o, r^k\right)\bar{H}^o = \bar{k}^w \qquad (7.11)$$

- clearing of the domestic home-good market in each country

$$\begin{aligned} \bar{\epsilon} &= \frac{p_s}{\lambda_s}\bar{y}_s && (7.12) \\ \bar{\epsilon}^o &= \frac{p_s^o}{\lambda_s}\bar{y}_s^o && (7.13) \end{aligned}$$

[2]The assumption that a sector employs the same technology as the corresponding sector in the other country eliminates the need to distinguish the cost and revenue functions for each country separately.

- and clearing of the world market for the agricultural good

$$\bar{\epsilon} + \bar{\epsilon}^o = \frac{p_a}{\lambda_a} \left(\bar{y}_a + \bar{y}_a^o \right) \tag{7.14}$$

For notational convenience we assume identical expenditure shares λ_j across countries.

This characterization bears a strong similarity to that of the basic three-sector model of Chapter 4. Major differences are the world capital market clearing condition (7.11) encompasses both countries, and the presence of a world market clearing condition for the agricultural good (7.14). This latter condition will require a third differential equation describing the trajectory of p_a to solve the model.

In the next two sections we follow a strategy similar to that pursued in the prior chapters. Our main objective in the first section is to obtain reduced form expressions for factor rental rates, and to derive output supply and expenditure functions. Then, in the second section, we use these reduced form expressions, the Euler condition and the world's flow budget constraint to reduce the system to three differential equations in arguments p_a, p_s, and world capital stock $\bar{k}^w$ per worker.

7.1.4 Reducing the dimensionality of the model

We begin, again, using the zero profit conditions (7.7) to obtain the home country's reduced form expression for its wage and capital rental rate as a function of home-good price, p_s. Likewise, use (7.8) to obtain the corresponding expressions for the other country. Express these results as

$$w = W(p_s) \tag{7.15}$$
$$r^k = R(p_s) \tag{7.16}$$

for the home country, and as

$$w^o = W(p_s^o) \tag{7.17}$$
$$r^k = R(p_s^o) \tag{7.18}$$

for the other country. Observe, we use the same functional representations for both countries because the countries only differ in their initial endowments which do not affect the zero profit conditions. Using the relationship $r^k = R(p_s) = R(p_s^o)$, define the level of p_s^o as a function of p_s

$$p_s^o = P^s(p_s)$$

Substitute this equation into expressions (7.17) and (7.18) to obtain

$$w^o = W^o(p_s) = W(P^s(p_s)) \tag{7.19}$$
$$r^k = R^o(p_s) = R(P^s(p_s)) \tag{7.20}$$

The system of zero profit conditions given by (7.7) and (7.8) yield the equilibrium wage and capital rental values (7.15) – (7.20), each of which are functions of the home country's non-traded good price.

We now use these reduced forms to further decrease the dimensionality of the model to a more manageable set of equations. First, substitute Equations (7.15), (7.16), (7.19), and (7.20) into the factor market clearing equations. These substitutions reduce the system to two labor market clearing conditions

$$\sum_{j=m,s} \tilde{C}_w^j(p_s)\,\bar{y}_j - \tilde{\pi}_w^a(p_a, p_s)\,\bar{H} = \ell \tag{7.21}$$
$$\sum_{j=m,s} \tilde{C}_{w^o}^{j,o}(p_s)\,\bar{y}_j^o - \tilde{\pi}_{w^o}^{a,o}(p_a, p_s)\,\bar{H}^o = (1-\ell) \tag{7.22}$$

and the capital market clearing condition

$$\sum_{j=m,s} \tilde{C}_{rk}^j(p_s)\,\bar{y}_j + \sum_{j=m,s} \tilde{C}_{rk}^{j,o}(p_s)\,\bar{y}_j^o -$$
$$\tilde{\pi}_{rk}^a(p_a, p_s)\,\bar{H} - \tilde{\pi}_{rk}^{a,o}(p_a, p_s)\,\bar{H}^o = \bar{k}^w \tag{7.23}$$

where

$$\tilde{C}_w^j(p_s) = C_w^j(W(p_s), R(p_s)), \;\; j = m, s$$

$$\tilde{\boldsymbol{\pi}}_w^a(p_a, p_s) = \boldsymbol{\pi}_w^a(p_a, W(p_s), R(p_s))$$

and

$$\tilde{C}_{w^o}^{j,o}(p_s) = C_{w^o}^j(W^o(p_s), R^o(p_s)), \ j = m, s$$

$$\tilde{\boldsymbol{\pi}}_{w^o}^{a,o}(p_a, p_s) = \boldsymbol{\pi}_{w^o}^a(p_a, W^o(p_s), R^o(p_s))$$

These three equations are linear in the four variables $\bar{y}_j$ and $\bar{y}_j^o$, $j = m, s$.

The fourth equation is derived from the two market clearing conditions for home-goods, and the market clearing condition for the agricultural good. Substitute into the world market clearing condition for the agricultural good, Equation (7.14), each country's expenditure implied by their respective home-good market clearing equations (7.12) and (7.13) to get

$$\frac{p_s}{\lambda_s}\bar{y}_s + \frac{P^s(p_s)}{\lambda_s}\bar{y}_s^o = \frac{p_a}{\lambda_a}\tilde{y}^A(p_a, p_s) \qquad (7.24)$$

where total agricultural supply per world worker in the two countries is given by

$$\begin{aligned}\tilde{y}^a(p_a, p_s) &= \boldsymbol{\pi}_{p_a}^a(p_a, W(p_s), R(p_s))\bar{H} \\ &\quad + \boldsymbol{\pi}_{p_a}^a(p_a, W^o(p_s), R^o(p_s))\bar{H}^o \qquad (7.25)\end{aligned}$$

This system is now reduced to the four equations (7.21) to (7.24) that are linear in outputs $\bar{y}_m$, $\bar{y}_s$, $\bar{y}_m^o$, and $\bar{y}_s^o$. Presuming, as we have throughout the text, an interior solution exists, this system of four equations can be solved to obtain the following output supply functions[3]

$$\tilde{y}^j\left(p_a, p_s, \bar{k}^w\right) = \bar{y}^j\left(p_a, p_s, \bar{k}^w, \bar{H}, \bar{H}^o\right), \ j = m, s \qquad (7.26)$$

$$\tilde{y}^{j,o}\left(p_a, p_s, \bar{k}^w\right) = \bar{y}^{j,o}\left(p_a, p_s, \bar{k}^w, \bar{H}, \bar{H}^o\right), \ j = m, s \qquad (7.27)$$

---

[3]An alternative derivation of the supply functions is to use the envelope properties of each economy's value added function

$$G(p_a, p_s, \bar{k}, \bar{H}) = W(p_s)\ell + R(p_s)\bar{k} + \boldsymbol{\pi}^a(p_a, W(p_s), R(p_s))\bar{H}$$

An advantage of this derivation is that supplies depend upon own country capital stock, and the properties of the supply functions are easily discernible from the properties of the value added function.

Then each country's equilibrium total expenditure can be obtained from the home-good market clearing condition

$$\tilde{\epsilon}\left(p_a, p_s, \bar{k}^w\right) = \frac{p_s}{\lambda_s}\tilde{y}^s\left(p_a, p_s, \bar{k}^w\right)$$

and

$$\tilde{\epsilon}^o\left(p_a, p_s, \bar{k}^w\right) = \frac{p_s}{\lambda_s}\tilde{y}^{o,s}\left(p_a, p_s, \bar{k}^w\right) \tag{7.28}$$

One can also derive equilibrium total world expenditures per world worker using the market clearing condition for the agricultural good

$$\tilde{\epsilon}^w\left(p_a, p_s\right) \equiv \frac{p_s}{\lambda_s}\tilde{y}^A\left(p_a, p_s\right) \tag{7.29}$$

For later reference, given a numerical solution to the model, the home country capital stock $\bar{k}$ can be calculated using the capital market clearing condition (7.23)

$$\sum_{j=m,s} \tilde{C}^m_{rk}\left(p_s\right)\bar{y}_j - \tilde{\pi}^a_{rk}\left(p_a, p_s\right)\bar{H} \tag{7.30}$$

where the supply functions for $\bar{y}_j$ are given by (7.26).

7.1.5 *Inter-temporal equilibrium*

Our main task is to solve for steady-state values of factor rental rates and home-good prices, and then derive three differential equations as functions of p_a, p_s,and $\bar{k}^w$. Given a numerical solution to these equations, we return to the intra-temporal equilibrium conditions for each country, and use the reduced forms derived in the previous section to calculate the values of the remaining endogenous variables. Capital flows, $\dot{d}-d\left(r^k - n - \delta\right)$, can be calculated as a "residual" from the balance of trade conditions (7.4) or (7.5), from which d can be calculated.

The steady-state equilibrium

The Euler condition (7.3) implies

$$r^{ss} = \rho$$

which, combined with expressions (7.15), (7.16), and (7.18), is sufficient information to identify the remaining steady-state wage rates and non-traded good prices

$$\left(p_s^{ss}, p_s^{o,ss}, w^{ss}, w^{o,ss}\right) \tag{7.31}$$

Our first target equation is the world budget constraint (7.6). We substitute into this equation the reduced from expressions for factor payments and total world expenditure (7.29) to get

$$\dot{\bar{k}}^w = W(p_s)\,\ell + W^o(p_s)(1-\ell) + \bar{k}^w\left(R(p_s) - n - \delta\right) + \tag{7.32}$$

$$\tilde{\pi}^a(p_a, p_s)\,\bar{H} + \tilde{\pi}^{a,o}(p_a, p_s)\,\bar{H}^o - \tilde{\epsilon}^w(p_a, p_s)$$

To obtain numerical values satisfying the steady-state equilibrium, an additional equation is required. We choose the world market clearing condition of the agricultural good (7.24). Substituting the home-good supply functions (7.26) and (7.27) into (7.24) yields

$$\frac{p_s}{\lambda_s}\tilde{y}^{s,o}\left(p_a, p_s, \bar{k}^w\right) + \frac{P^s(p_s)}{\lambda_s}\tilde{y}^s\left(p_a, p_s, \bar{k}^w\right) = \frac{p_a}{\lambda_a}\tilde{y}^A(p_a, p_s) \tag{7.33}$$

This system, (7.32) and (7.33), is linear in $\bar{k}^w$ but not linear in p_a. Given p_s^{ss} from (7.31), positive values of p_a and $\bar{k}^w$ satisfying this two equation system for $\dot{\bar{k}}^w = 0$ are steady-state values. To confirm this result, (7.30) can be used to calculate the steady-state value of home country $\bar{k}$. We can then return to the individual country budget constraints and, assuming $\dot{d}$ to be zero in the steady-state, calculate the value of d and assure that $\dot{\bar{k}}$, and $\dot{\bar{k}}^o$ are also equal to zero. Alternatively, the two-country budget constraints can be included along with (7.32) and (7.33) plus the constraint that $\bar{k}^w = \bar{k} + \bar{k}^o$. Given (7.31), the system is then composed of five equations in variables $\left\{p_a, \bar{k}^w, \bar{k}, \bar{k}^o, d\right\}$ for which a numerical solution should yield the same values as we obtain from working with the two equation system.

Deriving the transition path – differential equations

The derivation of the model's differential equations follows a pattern similar to that used in Chapter 4, except here we obtain an additional equation to account for the evolution of p_a. We begin by time differentiating expression (7.29)

$$\dot{\bar{\epsilon}} + \dot{\bar{\epsilon}}^o = \frac{d}{dt}\tilde{\epsilon}^w (p_a, p_s)$$

and use the Euler condition for each country to obtain

$$(\bar{\epsilon} + \bar{\epsilon}^o)(r - \rho) = \tilde{\epsilon}^w_{p_a}(p_a, p_s)\,\dot{p}_a + \tilde{\epsilon}^w_{p_s}(p_a, p_s)\,\dot{p}_s \tag{7.34}$$

We next need to replace the expenditure terms on the left hand side of (7.34). Although we can choose reduced form versions of the home-good market clearing conditions (7.12) and (7.13), in this case, a more convenient choice for total world expenditure is (7.29). Substitute $\bar{\epsilon} + \bar{\epsilon}^o = \tilde{\epsilon}^w (p_a, p_s)$ and replace $r = R(p_s) - \delta$ to get

$$\tilde{\epsilon}^w (p_a, p_s)\left[R(p_s) - \delta - \rho\right] = \tilde{\epsilon}^w_{pa}(p_a, p_s)\,\dot{p}_a + \tilde{\epsilon}^w_{ps}(p_a, p_s)\,\dot{p}_s \tag{7.35}$$

At this stage we have two equations in the three variables $\bar{k}^w$, p_a and p_s. We choose the home-good market clearing equation in the home country to complete the system. Use (7.28)

$$\dot{\bar{\epsilon}} = \frac{d}{dt}\tilde{\epsilon}\left(p_a, p_s, \bar{k}^w\right)$$

and the Euler condition to replace $\dot{\bar{\epsilon}}$ to obtain

$$\tilde{\epsilon}\left(p_a, p_s, \bar{k}^w\right)\left[R(p_s) - \rho - \delta\right] \tag{7.36}$$

$$= \tilde{\epsilon}_{pa}\left(p_a, p_s, \bar{k}^w\right)\dot{p}_a + \tilde{\epsilon}_{p_s}\left(p_a, p_s, \bar{k}^w\right)\dot{p}_s + \tilde{\epsilon}_{\bar{k}^w}\left(p_a, p_s, \bar{k}^w\right)\dot{\bar{k}}^w$$

The three equations (7.32), (7.35) and (7.36) are linear in $\dot{\bar{k}}^w$, $\dot{p}_a$ and $\dot{p}_s$ and, while complex, are easily expressed as a function of the level variables p_a, p_s and $\bar{k}^w$. Setting $\dot{p}_a$, $\dot{p}_s$ and $\dot{\bar{k}}$ equal to zero and numerically solving this system for $(p_a, p_s, \bar{k}^w)$ should yield the same numerical values for the steady-state as those obtained from solving (7.32) and (7.33) for p_a and $\bar{k}^w$.

A numerical solution to this system describes the transition path

$$\left\{p_a(t), p_s(t), \bar{k}^w(t)\right\}_{t\epsilon[0,\infty)}$$

from which factor payments can be calculated using (7.15), (7.16), (7.17), (7.18), (7.19), and (7.20), and the supplies using (7.25), (7.26), and (7.27). Sector resource employment can be calculated from (7.21), (7.22), and (7.23). Since $\dot{d}$ is zero in the steady-state, d can be calculated by recursion from the steady-state backward to the initial period.

7.2 A two-country world without capital mobility

The economic environment of the next modeled economy is identical to that of the former model except foreigners are not allowed to own capital assets employed in the other country. Consequently, it is not necessarily the case that the marginal value product of capital is equated across countries at each instant of time. Otherwise, we continue to assume the two countries employ the same technologies, hold identical preferences, and differ only in initial conditions. The notation also remains the same, i.e., a superscript "o" denotes the other country, variables not subscripted denote the home country, and the bar notation, e.g., "$\bar{y}$", indicates a variable that is expressed in per world worker terms.

7.2.1 Households and firms

The home country's household budget constraint in world worker terms is

$$\dot{\bar{k}} = w\ell + \bar{k}(r - n) + \pi\bar{H} - \bar{\epsilon} \tag{7.37}$$

while for the other country it is

$$\dot{\bar{k}}^o = w^o(1 - \ell) + \bar{k}^o(r^o - n) + \pi^o\bar{H}^o - \bar{\epsilon}^o \tag{7.38}$$

Removing foreign ownership of capital leads to two departures from the prior model: (i) there is no claim, d, and (ii) the returns to capital, r and r^o, although equal in the steady-state, are not necessarily equal over the transition path.

With unitary inter-temporal elasticity of substitution, the Euler condition for the home country is

$$\frac{\dot{\bar{\epsilon}}}{\bar{\epsilon}} = (r - \rho) \tag{7.39}$$

for the other country

$$\frac{\dot{\bar{\epsilon}}^o}{\bar{\epsilon}^o} = (r^o - \rho) \tag{7.40}$$

Assuming an identical time preference rate ρ in both regions leads to the result that, if a steady-state exists, each country's return to capital converges to the same steady-state value $r^{ss} = \rho$. The two Euler conditions which allow for different rates of return to capital imply this model has two state variables, $\bar{k}$ and $\bar{k}^o$.

7.2.2 Equilibrium

For a given sequence $\left\{\bar{k}(t), \bar{k}^o(t), \bar{\epsilon}(t), \bar{\epsilon}^o(t)\right\}_{t\epsilon[0,\infty)}$ of capital and expenditure pairs , and initial resource endowments

$$\{K(0), L(0), H, K^o(0), L^o(0), H^o\}$$

an intra-temporal equilibrium consists of the eleven-tuple sequence of positive values

$$\{w(t), w^o(t), r^k(t), r^{k,o}(t), p_s(t), p_s^o(t),$$
$$p_a(t), \bar{y}_m(t), \bar{y}_m^o(t), \bar{y}_s(t), \bar{y}_s^o(t)\}_{t\epsilon[0,\infty)}$$

satisfying the following eleven equations (7.41), (7.42), (7.43), (7.44), (7.45), (7.46), (7.47), (7.48), and (7.49) for each t

- zero profits in manufacturing and the home-good production,

$$C^m\left(w, r^k\right) = 1, \quad C^m\left(w^o, r^{k,o}\right) = 1 \tag{7.41}$$
$$C^s\left(w, r^k\right) = p_s, \quad C^s\left(w^o, r^{k,o}\right) = p_s^o \tag{7.42}$$

- labor market clearing in both countries

$$\sum_{j=m,s} C_w^m\left(w, r^k\right)\bar{y}_j - \boldsymbol{\pi}_w^a\left(p_a, w, r^k\right)\bar{H} = \ell \tag{7.43}$$
$$\sum_{j=m,s} C_{w^o}^m\left(w^o, r^{k,o}\right)\bar{y}_j^o - \boldsymbol{\pi}_{w^o}^a\left(p_a, w^o, r^{k,o}\right)\bar{H}^o = 1 - \ell \tag{7.44}$$

- capital market clearing in the home country

$$\sum_{j=m,s} C_{r^k}^m\left(w, r^k\right)\bar{y}_j - \boldsymbol{\pi}_{r^k}^a\left(p_a, w, r^k\right)\bar{H} = \bar{k} \tag{7.45}$$

- capital market clearing in the other country

$$\sum_{j=m,s} C_{r^{k,o}}^m\left(w^o, r^{k,o}\right)\bar{y}_j^o - \boldsymbol{\pi}_{r^{k,o}}^a\left(p_a, w^o, r^{k,o}\right)\bar{H}^o = \bar{k}^o \tag{7.46}$$

- clearing of the domestic home-good market in each country

$$\bar{\epsilon} = \frac{p_s}{\lambda_s}\bar{y}_s \tag{7.47}$$
$$\bar{\epsilon}^o = \frac{p_s^o}{\lambda_s}\bar{y}_s^o \tag{7.48}$$

- and clearing of the world market for the agricultural good

$$\bar{\epsilon} + \bar{\epsilon}^o = \frac{p_a}{\lambda_a}\left(\bar{y}_a + \bar{y}_a^o\right) \tag{7.49}$$

For notational convenience we continue to assume identical expenditure shares λ_j across countries.

The major differences in characterization of this model and the former is the additional endogenous variable, $r^{k,o}$, and the additional equation for clearing of the capital market in each country.

7.2.3 *Reducing the dimensionality of the model*

Following the same pattern as in our previous models for reducing dimensionality, we solve the zero profit conditions (7.41) and (7.42) for each country to obtain for the home country

$$
\begin{aligned}
w &= W(p_s) && (7.50)\\
r^k &= R(p_s) && (7.51)
\end{aligned}
$$

and the other country

$$
\begin{aligned}
r^{k,o} &= R(p_s^o) && (7.52)\\
w^o &= W(p_s^o) && (7.53)
\end{aligned}
$$

Note that although the technologies in each country are identical, in contrast to the previous model, without capital mobility between countries we cannot use the capital rental rate equations to express the price of the home-good p_s^o in the other country as a function of the home-good price in the home country.

Next, substitute (7.50) and (7.51) into the factor market clearing conditions (7.43) and (7.45). The result is a system of two factor market clearing conditions in five endogenous variables – $p_a, p_s, \bar{y}_m, \bar{y}_s$, and $\bar{k}$. These two equations can be used to solve for $\bar{y}_m$ and $\bar{y}_s$ as a function of p_a, p_s, and $\bar{k}$. Substituting (7.50) and (7.51) into $\pi^a(p_a, w, r)\bar{H}$ and applying Hotelling's lemma yields the home country's agricultural supply function. Let the resulting supply functions for the home country be denoted as

$$
\begin{aligned}
\tilde{y}^j\left(p_a, p_s, \bar{k}\right) &= \bar{y}^j\left(p_a, p_s, \bar{k}, \bar{H}\right), \; j = m, s && (7.54)\\
\bar{y}^a\left(p_a, p_s\right)\bar{H} &= \boldsymbol{\pi}^a_{p_a}\left(p_a, W(p_s), R(p_s)\right)\bar{H} && (7.55)
\end{aligned}
$$

Similarly, substitute (7.52) and (7.53) into the factor market clearing conditions (7.44) and (7.46). The result is a system of two factor market clearing conditions in five endogenous variables: $p_a, p_s^o, \bar{y}_m^o, \bar{y}_s^o$, and $\bar{k}^o$. These substitutions lead to the derivation of the supply functions for the other country:

$$
\begin{aligned}
\tilde{y}^{j,o}\left(p_a, p_s^o, \bar{k}^o\right) &= \bar{y}^{j,o}\left(p_a, p_s, \bar{k}^o, \bar{H}^o\right), \; j = m, s && (7.56)\\
\bar{y}^{a,o}\left(p_a, p_s^o\right)\bar{H}^o &= \boldsymbol{\pi}^a_{p_a}\left(p_a, W(p_s^o), R(p_s^o)\right)\bar{H}^o && (7.57)
\end{aligned}
$$

Equations (7.54), (7.55), (7.56), and (7.57) bear a strong similarity to the supply functions of Chapter 4, except the functions here are expressed in units of world workers rather than country workers.

We next draw upon the world market clearing condition of the agricultural good to express the capital stock in the other country as a function of the capital stock in the home country. Later, this linkage is used to allow us to express the model in terms of a single state variable. Substitute the home-good market clearing conditions (7.47) and (7.48) into (7.49) to obtain

$$\frac{p_s}{\lambda_s}\tilde{y}^s\left(p_a, p_s, \bar{k}\right) + \frac{p_s^o}{\lambda_s}\tilde{y}^{s,o}\left(p_a, p_s^o, \bar{k}^o\right) = \frac{p_a}{\lambda_a}\tilde{y}^A\left(p_a, p_s, p_s^o\right) \quad (7.58)$$

where $\tilde{y}^A\left(p_a, p_s, p_s^o\right)$ is total agricultural supply:

$$\tilde{y}^A\left(p_a, p_s, p_s^o\right) = \bar{y}^a\left(p_a, p_s\right)\bar{H} + \bar{y}^{a,o}\left(p_a, p_s^o\right)\bar{H}^o$$

Represent the home and other country expenditures in terms of their respective non-traded good supply

$$\tilde{\epsilon}\left(p_a, p_s, \bar{k}\right) = \frac{p_s}{\lambda_s}\tilde{y}^s\left(p_a, p_s, \bar{k}\right) \text{ and } \tilde{\epsilon}\left(p_a, p_s^o, \bar{k}^o\right) = \frac{p_s^o}{\lambda_s}\tilde{y}^s\left(p_a, p_s^o, \bar{k}^o\right) \quad (7.59)$$

Substituting expressions (7.50), (7.51), (7.52), and (7.53), and (7.59) into the flow the budget constraints (7.37) and (7.38), yield the reduced form flow budget constraints:

$$\dot{\bar{k}} = W\left(p_s\right)\ell + \bar{k}\left(R\left(p_s\right) - n - \delta\right) + \tilde{\boldsymbol{\pi}}^a\left(p_a, p_s\right)\bar{H} - \tilde{\epsilon}\left(p_a, p_s, \bar{k}\right) \quad (7.60)$$

and

$$\dot{\bar{k}}^o = W(p_s^o)\left(1 - \ell\right) + \bar{k}^o(R\left(p_s^o\right) - n - \delta) + \tilde{\boldsymbol{\pi}}^{a,o}\left(p_a, p_s^o\right)\bar{H}^o - \tilde{\epsilon}\left(p_a, p_s^o, \bar{k}^o\right) \quad (7.61)$$

where

$$\begin{aligned} \tilde{\boldsymbol{\pi}}^a\left(p_a, p_s\right)\bar{H} &= \boldsymbol{\pi}^a\left(p_a, W\left(p_s\right), R\left(p_s\right)\right)\bar{H} \\ \tilde{\boldsymbol{\pi}}^{a,o}\left(p_a, p_s^o\right)\bar{H}^o &= \boldsymbol{\pi}^a\left(p_a, W\left(p_s^o\right), R\left(p_s^o\right)\right)\bar{H}^o \end{aligned}$$

are reduced form expressions of the value-added by the sector specific factor in each country.

7.2.4 Inter-temporal equilibrium

Our task here is the same as the corresponding section of the previous model. We first solve for the steady-state values, and then derive the model's differential equations. In this case, there are five differential equations in variables $\{p_a, p_s, p_s^o, \bar{k}, \bar{k}^o\}$. This five equation system has two state variables $\bar{k}$ and $\bar{k}^o$. Each of the prior models contain a single state variable, which makes applying the time elimination method to obtain numerical solutions particularly straightforward. To maintain this numerical simplicity, we follow the suggestion of Mulligan and Sala-i-Martin (1991) and express one state variable as a function of the other. This substitution reduces the system to four differential equations in the four variables p_a, p_s, p_s^o, and $\bar{k}$.

The steady-state

To obtain the steady-state level of all endogenous variables, we first need the steady values for $\{w, w^o, r^k, r^{k,o}, p_s, p_s^o, p_a, \bar{k}, \bar{k}^o\}$. Once obtained, the reduced form equations derived in the previous section permit recovering the steady-state values of the remaining variables.

If a steady-state exists, the Euler conditions (7.39) and (7.40) yield the steady-state level of r and r^o, which in this case are both equal to ρ. Given $r^{ss} = \rho$, calculate the steady-state wage and home-good price of each country using the reduced form factor payment equations (7.50), (7.51), (7.52), and (7.53).

The three remaining variables are the price of the agricultural good p_a and capital stock per world worker in each country, $\bar{k}$ and $\bar{k}^o$. To identify the level of these three variables, we need a system of three independent equations. In this case, consider the agricultural good's world market clearing equation, and the home and other country's reduced form budget constraints, Equations (7.58), (7.60), and (7.61). If a steady-state exists, the latter two equations satisfy $\dot{\bar{k}} = \dot{\bar{k}}^o = 0$. Substituting the steady-state values of p_s and p_s^o into these three equations yields a system of three equations in p_a, $\bar{k}$, and $\bar{k}^o$. These equations are linear in the capital stock variables but not in p_a. The numerical values $\{p_a^{ss}, \bar{k}^{ss}, \bar{k}^{o,ss}\}$ satisfying these equations permit the calculation of the steady-state values of the model's remaining endogenous variables.

Deriving the transition path – differential equations

The derivation of the model's differential equations, while somewhat more involved than the previous models, nevertheless follows the same pattern of drawing upon the Euler conditions, the market clearing conditions for home-goods and the agricultural good, and each country's budget constraint. We proceed first with a derivation that results in five differential equations in variables $\{p_a, p_s, p_s^o, \bar{k}, \bar{k}^o\}$ where $\bar{k}$ and $\bar{k}^o$ are the two state variables. We then express one state variable as a function of the other which reduces the system to four differential equations in the four variables $\{p_a, p_s, p_s^o, \bar{k}\}$.

We begin with the Euler condition (7.39) and the home-good market clearing condition (7.59) for the home country. Time differentiate the home-good market clearing condition (7.59) to get

$$\begin{aligned}\dot{\bar{\epsilon}} &= \frac{d}{dt}\tilde{\epsilon}\left(p_a, p_s, \bar{k}\right) = \\ &\tilde{\epsilon}_{p_a}\left(p_a, p_s, \bar{k}\right)\dot{p}_a + \tilde{\epsilon}_{p_s}\left(p_a, p_s, \bar{k}\right)\dot{p}_s + \tilde{\epsilon}_{\bar{k}}\left(p_a, p_s, \bar{k}\right)\dot{\bar{k}},\end{aligned}$$

Then, replace $\dot{\bar{\epsilon}}$ with the Euler condition (7.39), and $\bar{\epsilon}$ by (7.59) to obtain

$$\begin{aligned}&\tilde{\epsilon}\left(p_a, p_s, \bar{k}\right)\left[R\left(p_s\right) - \delta - \rho\right] = \\ &\tilde{\epsilon}_{p_a}\left(p_a, p_s, \bar{k}\right)\dot{p}_a + \tilde{\epsilon}_{p_s}\left(p_a, p_s, \bar{k}\right)\dot{p}_s + \tilde{\epsilon}_{\bar{k}}\left(p_a, p_s, \bar{k}\right)\dot{\bar{k}} \qquad (7.62)\end{aligned}$$

Perform the same derivation for the other country using Equations (7.40) and (7.59). Express the result as

$$\begin{aligned}&\tilde{\epsilon}^o\left(p_a, p_s^o, \bar{k}^o\right)\left[R\left(p_s^o\right) - \delta - \rho\right] = \\ &\tilde{\epsilon}_{p_a}^o\left(p_a, p_s^o, \bar{k}^o\right)\dot{p}_a + \tilde{\epsilon}_{p_s^o}\left(p_a, p_s^o, \bar{k}^o\right)\dot{p}_s^o + \tilde{\epsilon}_{\bar{k}^o}^o\left(p_a, p_s^o, \bar{k}^o\right)\dot{\bar{k}}^o \quad (7.63)\end{aligned}$$

We now focus on the market clearing condition for the agricultural good, and begin with (7.49). As implied above, to simplify the derivation the household expenditure shares λ_a on the agricultural good are assumed to be identical in both countries. Substitute into (7.49) the supply functions (7.55) and (7.57), and time differentiate to obtain

$$\dot{\bar{\epsilon}} + \dot{\bar{\epsilon}}^o = \frac{d}{dt}\left(\frac{p_a}{\lambda_a}\tilde{y}^A\left(p_a, p_s, p_s^o\right)\right)$$

Replace $\dot{\bar{\epsilon}}$ and $\dot{\bar{\epsilon}}^o$ by the corresponding Euler conditions, and express the result as

$$\tilde{\epsilon}\left(p_a, p_s, \bar{k}\right)\left(R\left(p_s\right) - \delta - \rho\right) + \tilde{\epsilon}^o\left(p_a, p_s^o, \bar{k}^o\right)\left(R\left(p_s^o\right) - \delta - \rho\right) = \tag{7.64}$$
$$\frac{1}{\lambda_a}\left(\tilde{y}^A\left(p_a, p_s, p_s^o\right) + p_a\tilde{y}^A_{p_a}\left(p_a, p_s, p_s^o\right)\dot{p}_a\right) +$$
$$\frac{1}{\lambda_a}\left(p_a\tilde{y}^A_{p_s}\left(p_a, p_s, p_s^o\right)\dot{p}_s + p_a\tilde{y}^A_{p_s^o}\left(p_a, p_s, p_s^o\right)\dot{p}_s^o\right)$$

At this stage, we have the three equations (7.62), (7.63) and (7.64) in level variables $\left\{p_a, p_s, p_s^o, \bar{k}, \bar{k}^o\right\}$. The two budget constraints (7.60) and (7.61) complete the system. This system is linear in the rate of change variables $\left\{\dot{p}_a, \dot{p}_s, \dot{p}_s^o, \dot{\bar{k}}, \dot{\bar{k}}^o\right\}$ and hence easily expressed in terms of the level variables. However, the two state variables, $\bar{k}$ and $\bar{k}^o$, make applying the time elimination method to empirically solve this system problematic.[4]

For the case of this model, the two state variable problem is easily resolved. The market clearing equation for the agricultural good (7.58) is linear in $\bar{k}$ and $\bar{k}^o$. Thus, we can use this equation to express the capital stock of the other country as a function of the capital stock in the home country. Denote this relation as

$$\bar{k}^o = \tilde{k}\left(p_a, p_s, p_s^o, \bar{k}\right) \tag{7.65}$$

---

[4]This system, if well behaved, will tend to produce two unequal and negative eigenvalues in the neighborhood of the steady state which can lead to a transition path of the model's endogenous variables that are not consistent with initial conditions.

Next, substitute (7.65) into the other country's budget constraint, (7.61) to obtain the rather lengthy expression

$$\dot{\bar{k}}^o = W\left(p_s^o\right)(1-\ell) + \tilde{k}\left(p_a, p_s, p_s^o, \bar{k}\right)\left[R\left(p_s^o\right) - \delta - n\right] + \tag{7.66}$$
$$\tilde{\pi}^{a,o}\left(p_a, p_s^o\right)\bar{H}^o - \tilde{\epsilon}^o\left(p_a, p_s^o, \tilde{k}\left(p_a, p_s, p_s^o, \bar{k}\right)\right)$$

Similarly, substitute (7.65) into expressions (7.63) and (7.64). After making these substitutions, however, note that $\dot{\bar{k}}^o$ remains an argument in the updated versions of Equations (7.63) and (7.64). To resolve this problem, simply substitute expression (7.66) into the updated expressions. Doing so yields a system of four equations in the four level variables $\left\{p_a, p_s, p_s^o, \bar{k}\right\}$. The equations also remain linear in $\left\{\dot{p}_a, \dot{p}_s, \dot{p}_s^o, \dot{\bar{k}}\right\}$. One can now readily apply the time elimination method.

Upon obtaining the numerical solution

$$\left\{p_a^*(t), p_s^*(t), p_s^{o*}(t), \bar{k}^*(t)\right\}_{t\epsilon[0,\infty)}$$

$\bar{k}^{o*}(t)$ is recovered from (7.65). The remaining variables

$$\left\{w^*(t), w^{o*}(t), r^{k*}(t), r^{k,o*}(t)\right\}_{t\epsilon[0,\infty)}$$

are calculated from the reduced forms (7.50), (7.51), (7.52), and (7.53), which then permits the calculation of all of the remaining endogenous variables of the model.

7.3 Numerical examples

In this section we present two numerical examples. The first example considers a two-country world model with capital mobility across countries. The second example prohibits the citizens of one country from holding capital stock employed in the other. The structure of the Turkish economy in both examples is virtually identical to the empirical example presented in Chapter 4. The rest of the world (referred to as the other region) is based mostly on hypothetical data and constructed so the economies of the two regions employ the same technologies and otherwise are almost identical on a GDP per worker basis. In 2001, aggregate Turkish GDP is \$135.6 billion in 2001 U.S. dollars, while the corresponding GDP of the other country is about 6.5 times larger at \$880 billion. On a per worker basis, Turkey's GDP is \$5,643 and the other region's GDP per worker is \$5,708. The other region is endowed with more land, so its land rental payment as a percent of aggregate GDP is 1.6 percent, while for Turkey, this value is 0.9 percent. The initial capital to GDP ratio is identical in both regions, which in turn, is equal to the initial capital to GDP ratio in each of our previous numerical examples. Under this setup, Turkey exports the industrial good and imports the agricultural good in both models. The intent is to facilitate comparison of the results obtained here with the results obtained for the empirical three-sector model of Chapter 4.

The parameter values, $(\theta,\ \rho,\ \delta,\ x,\ n,\ \alpha,\ \beta,\ \phi_l,\ \phi_k)$, reported in Table 4.1 remain unchanged for both empirical models of this chapter, and are identical for both regions. For both regions, we assume the household expenditure share of $\lambda_a = 0.227$ for the agricultural good which is a larger value than the value reported in Table 4.1.

7.3.1 The capital mobility model

An important difference between the two regions is that Turkey incurs a trade surplus in the initial period equal to $rD - \dot{D} =$ \$10.9 billion (see Equation 7.1) while the other region incurs a trade deficit. Turkey's trade surplus is 16.7 percent of her total capital stock earning in 2001 of \$65.6 billion (or 8.1 percent of GDP) which is remunerated to the other region's households. Consequently, in this example, total household income per Turkish worker in the initial period is 6,518 million 2001 Lira per annum.

The model predicts a rate of growth of GDP per Turkish worker of 3.3 percent per annum in year 2002, and reaches the half-way point to the steady-state rate of growth per worker by year 2012 of 2.6 percent. The time required to double GDP per worker is about 26 years. By the year 2050, Turkey is within 3.4 percent of its long-run rate of growth in GDP per worker of 1.9 percent. All prior examples required 27–29 years to double income. Capital outflows from the other region to Turkey are tending to speed up the country's transition to long-run growth. From Table 7.1 the capital stock to GDP ratio rises from an initial value of about 4.1 to about 4.8 by year 2031. At the half-way point, GDP per worker has increased by a factor of 1.4 of its base 2001 level and by a factor of 2.15 by 2031. As the economy grows, the ratio of GNP to GDP falls so that Turkish households are increasing their share of GDP.

The higher level of capital stock per Turkish worker compared to the levels shown in Table 4.1 partially explain the higher level of wage income in this example. While total capital earnings are higher here than shown in Table 4.1, capital earnings to Turkish households are marginally less than that example due to the share of Turkish capital stock held by households in the other region. Real expenditure per worker also exceeds the expenditure levels reported in Table 4.1.

A notable departure from the previous examples is the increase in land rental income, starting from a lower base value in 2001 and exceeding the 2031 land rental values predicted by all prior models. This result obtains because preferences are homothetic, and agriculture is the most labor intensive sector which causes the price of this good to increase as incomes grow and capital deepening occurs. The lower initial land rental value is due to a marginally higher initial total capital stock in this economy. This higher level of capital tends to increase slightly the wage rate, which in turn penalizes the labor intensive agricultural sector. Turkish agriculture also faces import competition from the other country. The share of GDP accounted for by wages and capital remain relatively constant over time, at about 51 and 48 percent, respectively.

The effects of capital deepening on changes in sectoral value shares in GDP, and on the sectoral shares of labor and capital

Table 7.1 Factor income and expenditure in millions of 2001 Turkish Lira

Year	GDP per worker	GNP per worker	Capital per worker	Foreign capital per worker	Wage income per worker	Total capital earnings per worker	Capital earnings to other country	Land rental income per worker	Expend-iture per worker	Saving to GDP
2001	6943	6518	28540	3410	3339	3556	425	48	3983	0.39
2006	8135	7651	35098	4089	3921	4157	484	57	4778	0.38
2011	9356	8809	41763	4783	4517	4772	546	67	5587	0.37
2016	10623	10010	48589	5498	5134	5411	612	77	6419	0.36
2021	11954	11272	55650	6243	5783	6084	683	87	7287	0.35
2026	13368	12610	63030	7027	6471	6799	758	98	8200	0.35
2031	14881	14042	70822	7861	7207	7565	840	110	9171	0.35

Source: model results

are shown in Table 7.2. For both Turkey and the other region, the evolution of sector value shares is similar to the closed two-sector model of Chapter 3. The pattern of the evolution of other region's sector shares in GDP are similar to those of Turkey, but depart somewhat due to foreign trade. Since the other region exports the agricultural good and imports the industrial good, its agriculture's share is higher and the industrial sector share is marginally smaller than that reported for Turkey. In the long-run, agriculture and service sectoral value shares in GDP approach from below the value of consumption shares and remain higher for the industrial sector to account for capital depreciation and foreign trade. The decline in industry's share of GDP can be attributed to the enhanced ability of agriculture to compete for resources (compared to the basic three-sector model). We discuss the impact of agriculture on industrial growth shortly. Throughout transition growth Turkey's industrial exports exceed the value of agricultural imports, with the difference equal to the value of capital rental payments to foreign owners of capital employed in Turkey plus net new capital flows each period.

The rise in the price of the agricultural good, for reasons discussed below, induces a supply response causing the share of workers and capital employed in the sector to rise over time. This adjustment is needed to meet the growth in world demand caused by rising incomes. This result is in contrast to the other examples in which agriculture decreased the share of the economy's resources employed in the sector. The international flow of capital speeds up the capital deepening process for all sectors of the Turkish economy. Agriculture's capital to labor ratio increases by a factor of 1.23 between 2001 and 2004, and by a factor of 2.52 over the 2001 to 2031 period. This level of capital deepening compares to 1.2 and 2.4 for comparable periods of the three-sector model of Chapter 4. This modestly higher level of capital per worker over time, as well as the rise in the price of the relatively labor intensive agricultural good, accounts for the higher annual wage income compared to example of Chapter 4.

Table 7.2 Sector value shares in GDP and sector factor shares in total factors, Turkey

	Sector Share in GDP			Labor Share in			Capital Share in		
Year	Industry	Agriculture	Service	Industry	Agriculture	Service	Industry	Agriculture	Service
2001	0.529	0.086	0.385	0.480	0.097	0.423	0.582	0.064	0.354
2006	0.518	0.088	0.394	0.468	0.099	0.432	0.572	0.066	0.363
2011	0.510	0.090	0.400	0.460	0.101	0.439	0.564	0.067	0.369
2016	0.504	0.091	0.405	0.455	0.102	0.444	0.558	0.068	0.374
2021	0.500	0.092	0.409	0.450	0.103	0.447	0.553	0.068	0.378
2026	0.496	0.092	0.411	0.447	0.103	0.450	0.550	0.069	0.381
2031	0.494	0.093	0.413	0.445	0.104	0.452	0.548	0.069	0.383

Source: model results

Table 7.3 Growth in industrial output and factor contributions, Turkey

		Contributions to Growth		
Year	Growth in output	Combined price effect	Capital effect	Effective labor
2001	0.0438	−0.1363	0.3312	−0.1512
2006	0.0405	−0.0975	0.2927	−0.1547
2011	0.0384	−0.0704	0.2662	−0.1573
2016	0.0370	−0.0512	0.2476	−0.1593
2021	0.0361	−0.0375	0.2343	−0.1608
2026	0.0354	−0.0275	0.2248	−0.1619
2031	0.0349	−0.0202	0.2178	−0.1627

Source: Model results

The contributions to sector growth are shown in Table 7.3, 7.4, and 7.5. Growth in agricultural and service output are higher during the earlier periods of transition than is the case for all preceding examples, while growth of industrial output is marginally lower. Rybczynski-like effects of capital deepening on the relatively capital intensive industrial sector are evident (Table 7.3), but the magnitudes are smaller than the corresponding magnitudes reported for this sector in the preceding chapters. The combined negative price effects on output growth from the increase in agricultural and service good prices lead to a negative contribution to growth that is marginally less negative than that found in preceding examples. These more moderate effects on industrial output growth are due to the increased competition for resources from the agricultural and service sectors.

As noted above, in contrast to the transition dynamics for agriculture in the prior examples, here agriculture converges towards its long-run rate of output growth from above. The positive combined price effect implies the positive effect associated with agriculture's increasing price dominates the negative effect on agricultural output from the increasing service good price. The percentage increase in agriculture's price is twice that of the service good price, hence agriculture experiences a positive external and internal terms of trade effect. In both regions, with land as a fixed factor of production, agriculture experiences diminishing returns to labor and capital. This causes the world market for agriculture to clear at a price that increases at a higher rate than is the case of the service sector price in either country.

Table 7.4 Growth in agricultural output and factor contributions, Turkey

		Contributions to Growth			
Year	Growth in output	Price effect	Wage effect	Interest rate effect	Technical change
2001	0.0488	0.0636	−0.1060	0.0576	0.0336
2006	0.0443	0.0445	−0.0743	0.0404	0.0336
2011	0.0412	0.0317	−0.0528	0.0287	0.0336
2016	0.0391	0.0228	−0.0380	0.0207	0.0336
2021	0.0376	0.0165	−0.0275	0.0150	0.0336
2026	0.0365	0.0120	−0.0201	0.0109	0.0336
2031	0.0357	0.0088	−0.0147	0.0080	0.0336

Source: Model results

The marginally higher rate of capital deepening, as revealed when comparing the capital to labor ratio of Table 7.2 with the same ratio reported in Table 4.2, causes income from wages and land rental rates to increase, which increases the demand for final goods at a marginally higher rate than that which occurred in the previous examples. The negative effect of increased wages on the cost of producing the agricultural good dominates the positive effect of the decline in the rental rate of capital. These economic forces leave the rising agricultural price as one of the driving incentives to increase agricultural output and clear the world market for agricultural goods.

Relative to the prior examples, the slightly higher rate of growth in GDP per worker is also reflected in the marginally higher rate of growth in service sector output, particularly during the earlier years of transition. Between 2001 and 2016, the negative effects of capital deepening – which exceed the same effects for the example in Chapter 4 – are not compensated for by the positive effects of growth in labor services. The price of the service good must rise to compete for the labor and capital necessary to clear the domestic market.

Table 7.5 Growth in service sector output and factor contributions, Turkey

		Contributions to Growth		
Year	Growth in output	Combined price effect	Capital effect	Effective labor
2001	0.0517	0.1840	−0.3753	0.2429
2006	0.0463	0.1258	−0.3171	0.2376
2011	0.0427	0.0880	−0.2792	0.2339
2016	0.0401	0.0625	−0.2536	0.2313
2021	0.0383	0.0449	−0.2359	0.2294
2026	0.0371	0.0325	−0.2234	0.2280
2031	0.0361	0.0236	−0.2145	0.2270

Source: Model results

In developing the analytical framework, we assumed the technology in both countries are identical to facilitate notation. Identical technologies led to factor price equalization. Since land does not enter the wage and capital rental rate equations (7.15), (7.16), (7.17), and (7.18), factor price equalization and equalization the price of the service good also obtain. A shock to the scale parameter of the service good production function in the other region will alter these results by changing the domestic terms of trade in the other region with effects on the home country.

7.3.2 Model without capital mobility

In this senario, the structure of the Turkish economy and that of the other region remain unchanged from the example above with two exceptions: (i) the capital stock owned by the other region just prior to the initial period, which is responsible for generating most of the approximately \$10.9 billion capital flow in the initial period, is now considered the property of Turkish households, and (ii) capital flow between the two regions is prohibited. Otherwise the two empirical examples are identical. Each region's value of exports equal the value of imports. Turkey continues to import the agricultural good, but the value of exports plus imports now only account for about 10 percent of GDP. The unit cost $\mathcal{E}(p_a, p_s)$ of consuming the aggregrate good, q, is slightly higher in both regions during the earlier phase of transition growth than in the preceding mobile capital example. This setup leads to the same steady-state values for wages, the capital rental rate and the level of capital stock as obtained for the mobile capital model. The pattern of adjustment over time is also similar. We thus focus discussion on the dissimilarities between the two results.

Factor income and expenditure per worker are reported in Table 7.6. Since there is no remuneration to foreign owners of Turkish capital, GNP per worker is equal to GDP per worker, and it is higher than the GNP values for corresponding years reported in Table 7.1. The stock of capital per worker is lower and saving to GDP is higher here than those reported above. The lower capital per worker results in a modestly lower initial wage, lower capital earnings and lower returns to land.

Thus, the absence of capital flows causes the economy's total returns to labor and land to be modestly lower than in the previous example. Because our setup "confiscates" the capital of the other country just prior to the initial period, there is no remuneration to foreign holders of domestic capital stock so that domestic households are better off than in the preceding example. The rate of growth in GDP and time to double income are almost identical to the corresponding values reported above.

In contrast to the preceding example, the lower rate of growth in the country's capital stock is coupled with moderately higher expenditure levels (since domestic households now retain all of the factor earnings), and the service sector employs a larger share of the economy's labor and capital. To attract the necessary resources for market clearing, the world price of the agricultural good increases by about 8 percent of its base period value by 2031, compared to 7 percent in the previous example.

Table 7.6 Factor income and expenditure in millions of 2001 Turkish Lira

Year	GDP per worker	Capital per worker	Wage income per worker	Capital earnings per worker	Land rental income per worker	Expenditure per worker	Saving to GDP
2001	6864	27977	3304	3514	45	4027	0.41
2006	8065	34568	3893	4117	55	4850	0.40
2011	9292	41262	4493	4734	65	5687	0.39
2016	10564	48110	5115	5374	76	6547	0.38
2021	11899	55185	5767	6046	86	7442	0.37
2026	13315	62572	6458	6760	97	8383	0.37
2031	14829	70362	7196	7524	109	9382	0.37

Source: model results

The price of the service good rises by 4 percent of its base period value by 2031 compared to 3 percent in the previous example. Consequently, both industry and agriculture account for a smaller value share of GDP, and their production levels are marginally lower than those in the previous example. We thus have the result that the lack of capital flows in transition growth has decreased the two regions' value of trade in GDP, i.e., lessened the extent of the specialization and division of labor, and increased the share of the economy's total resources employed in the production of the non-traded good (Table 7.7).

Price and Rybczynski-like effects on growth in sectoral output are reported in Table 7.8, 7.9, and 7.10 where it can be seen that growth in industrial sector output is slightly less than that of the previous example. The slower rate of growth in the stock of capital leads to a slightly less negative effect on service sector output growth. In this case, the higher level of expenditures and demand for the service good induce larger negative terms of trade effects on industrial sector output growth (see Table 7.8). Prohibiting inter-regional capital flows lessens the degree to which Turkey (the other region) can specialize in the production of the industrial (agricultural) good during the earlier phases of transition growth. In this case, the world price of the agricultural good, in terms of the industrial good as numeraire, rises more rapidly in the early phase of transition growth than it did in the preceding example. This effect helps to counter the negative effect on the cost producers face from the rise in wages (Table 7.9b).

In contrast to the preceding example, to clear the domestic market for the service good, the growth in output of the service sector during the earlier phase of transition is higher in Turkey than it is in the other region, and higher than the preceding example. The combined price effect and the effect of growth in the services of labor are larger than the same effects in the preceding example while the slower growth in the country's stock of capital causes a less negative Rybczynski-like effect on output growth.

Table 7.7 Sector value shares in GDP and sector factor shares in total factors, Turkey

	Sector Share in GDP			Labor Share in			Capital Share in		
Year	Industry	Agriculture	Service	Industry	Agriculture	Service	Industry	Agriculture	Service
2001	0.524	0.083	0.393	0.474	0.093	0.432	0.577	0.062	0.362
2006	0.511	0.086	0.403	0.461	0.096	0.442	0.564	0.064	0.372
2011	0.501	0.088	0.410	0.452	0.099	0.449	0.555	0.066	0.379
2016	0.495	0.090	0.416	0.445	0.100	0.454	0.548	0.067	0.385
2021	0.490	0.091	0.419	0.441	0.101	0.458	0.543	0.068	0.389
2026	0.486	0.092	0.422	0.437	0.102	0.461	0.540	0.069	0.391
2031	0.483	0.092	0.424	0.434	0.103	0.463	0.537	0.069	0.394

Source: Model results

Table 7.8 Growth in industrial output and factor contributions, Turkey

Year	Growth in output	Contributions to Growth: Combined price effect	Capital effect	Effective labor
2001	0.0436	−0.1443	0.3410	−0.1530
2006	0.0404	−0.1033	0.3009	−0.1571
2011	0.0383	−0.0747	0.2733	−0.1603
2016	0.0369	−0.0544	0.2540	−0.1626
2021	0.0360	−0.0398	0.2402	−0.1644
2026	0.0353	−0.0292	0.2302	−0.1657
2031	0.0348	−0.0214	0.2230	−0.1667

Source: Model results

Table 7.9 Growth in agricultural output and factor contributions, Turkey

Year	Growth in output	Contributions to Growth: Price effect	Wage effect	Interest rate effect	Technical change
2001	0.0520	0.0692	−0.1112	0.0605	0.0336
2006	0.0464	0.0482	−0.0777	0.0422	0.0336
2011	0.0426	0.0342	−0.0552	0.0300	0.0336
2016	0.0401	0.0246	−0.0397	0.0216	0.0336
2021	0.0383	0.0178	−0.0287	0.0156	0.0336
2026	0.0370	0.0129	−0.0209	0.0114	0.0336
2031	0.0361	0.0095	−0.0153	0.0083	0.0336

Source: Model results

7.4 Conclusions

In the empirical models of Chapters 3, 4, 5, and 6, the prices of traded goods were stationary and international capital flows were prohibited. The main motivation for this chapter was to relax these assumptions. Since policy interventions, institutional constraints and other capital market barriers can constrain or

Table 7.10 Growth in service sector output and factor contributions, Turkey

		Contributions to Growth		
Year	Growth in output	Combined price effect	Capital effect	Effective labor
2001	0.0526	0.1882	−0.3739	0.2382
2006	0.0469	0.1281	−0.3139	0.2327
2011	0.0431	0.0894	−0.2751	0.2288
2016	0.0404	0.0634	−0.2490	0.2261
2021	0.0385	0.0454	−0.2310	0.2241
2026	0.0372	0.0329	−0.2183	0.2227
2031	0.0362	0.0239	−0.2093	0.2217

Source: Model results

restrict capital flows, we also examined the case where traded good prices were endogenous, but international capital flows were prohibited.

We drew upon the three-sector small open economy model of Chapter 4 and added another region which we treated as the rest of the world. The result is a model that can be seen as a relatively straightforward extension of the simpler single country model. To avoid confusion that can arise by doubling the number of variables, we assumed the countries employed the same technologies and held identical preferences. The addition of the other region caused the price of the agricultural good to become endogenous, while the price of the manufacturing good was the numeraire. The first model presented allows for household in one region to own capital stock employed in the other region. We refer to this as the mobile capital model. The second model prohibits capital flows. Analytically, the mobile capital model's intra-temporal equilibrium featured a single capital market clearing equation in which the sectoral demand for capital in both regions equaled world capital supply. The model's inter-temporal conditions entailed three differential equations, two of which are policy functions for the control variables p_a and p_s, and one state variable equation for the world's stock of capital.

The model without capital mobility featured two intra-temporal capital market clearing conditions, one for each country. In this case, depending upon initial conditions, factor price equalization need not occur in transition growth. In principle, the model has two state variables. In order to maintain numerical consistency for solving all of the models presented, we used the world market clearing condition for the agricultural good to help express the capital stock of the rest of the world as a function of the stock of capital in the home country. This reduces the inter-temporal equilibrium to four differential equations, three of which are policy functions for the agricultural price and the two home-good prices. The capital stock in the home country becomes the state variable. A numerical solution to this system allows for the derivation of all of the remaining endogenous variables.

The chapter concludes with two numerical examples, one for each model. To facilitate comparison with the other empirical examples of Chapters 4, 5, and 6, the Turkish economy remains virtually identical to the economy modeled in Chapter 4. The other region is based mostly on hypothetical data and presumed to employ the same technology as employed in the Turkish economy. While the other region's GDP is about 6.5 times that of Turkey, its GDP per worker is only marginally higher, and its endowment of land is greater so that Turkey imports the agricultural good and exports the industrial good. Initially, Turkey's trade surplus is about 16.7 percent of her total earnings from capital stock. A highlight of the results is that capital flows increase the rate of transition growth in early periods compared to the other examples, but an undesirable feature common to both examples is the rise in the price of the agricultural good which is not consistent with historical observation. This suggests the need to consider a preference structure which induces households to decrease their share of expenditure on the agricultural good as incomes rise.

The second example was set up as though the domestic capital held by households in the other country was "confiscated" and became the property of Turkish households. Otherwise the data are common to the two examples. No capital flows caused a slightly slower rate of transition growth than growth in the mobile capital model for the same time periods, a smaller share of GDP was traded and, consequently, the economy experienced smaller gains from specialization that capital flows could otherwise induce.

8

Data Issues and the Social Accounting Matrix

8.1 Introduction

The empirical implementation of general equilibrium models is greatly facilitated by the organization of input-output data, and related data on national income, production and public accounts into the framework known as a social accounting matrix (SAM). A SAM is a double-entry accounting system that owes its origins to the work of Nobel Laureate, Sir Richard Stone, founder of the United Nations' System of National Accounts (SNA).[1] A SAM shows the major flows of income sources and expenditures of an economy over a specific time period, usually one year. It shows the major economic transactions among the various agents of an economy, for the given period of time. Furthermore, as shown later, the SAM provides a direct link between the theoretical dynamic general equilibrium models in Chapters 3, 4, 5, and 6, and their respective empirical counterparts.

A SAM is a square matrix that provides a snapshot of the economic activity of a country, region, or regions over a period of time. Each row in the matrix represents an "account" that has a matching column account. For example, the row "Labor" in Table 8.1 has a corresponding column "Labor," and the row "Activity 1" has a corresponding column "Activity 1." Each cell in the SAM represents the payment from the account of its column to the account of its row; e.g., the intersection of accounts "Activity 1" and "Labor" in Table 8.1, is the payment from activity 1 to labor. The sum of entries along a row account gives the total receipts for that account, while the sum of entries down a column account gives the total expenditures of that account.

[1]See Robinson (1989) for a good discussion of social accounting matrices.

T.L. Roe et al., *Multisector Growth Models*, DOI 10.1007/978-0-387-77358-2_8,

The double-entry bookkeeping feature of the SAM requires the row sum of an account to be equal to its corresponding column sum. Hence, a necessary condition for Table 8.1 to represent a correctly constructed SAM is for the following equalities to hold: $\text{Cost}_j = \text{Production}_j$, $\text{Supply}_j = \text{Demand}_j$, $j = 1, 2$; Lab Pay = Lab Inc, Cap Pay = Cap Inc; and Expenditure = Income, and Savings = Investment.

To date, a universal guideline for designing a SAM does not exist, as the desired features of a SAM will vary with the economic issues one is analyzing. For example, if studying the economic growth of a country with and without indirect taxes on the manufacturing and agricultural sectors, a SAM with the "activities" agriculture, manufacturing, services, and "inputs" capital, labor, and land might meet the modeler's needs. On the other hand, if interested in the impact a pernicious disease might have on an agricultural based rural economy the modeler might want to have several sub-accounts for agriculture and then an account for non-agricultural production. Hence, the analyst will design a SAM according to the needs of the modeling exercise.

For the analyst linking theory to macroeconomic data, the SAM serves two major purposes. First, it is used to identify the parameters of the primitives of the underlying economy – the sectoral production technologies and household preferences. Second, it is used to help ensure the theoretical model and its parameterization is logically consistent. In this chapter we focus attention on the process of linking the theoretical models in this book to their empirical counterparts. In other words, here we discuss our approach to calculating the value of each parameter used in the empirical models presented in Chapters 3, 4, 5, and 6: e.g., the value of sectoral technology production cost shares and scale parameters like α, β, and Ψ_s, and the value of consumption shares λ_a and λ_m^g.

Several parameter values, like θ, ρ, and δ, or the elasticity of factor substitution for constant elasticity of substitution production technologies, have no direct relationship with a SAM. The choice of these values can, however, influence variables in the model through their potential influence on the size of the initial capital stock, and hence, on the initial rate of return to capital. This point comes up in Section 8.2, where we discuss approaches to estimating the initial capital stock. Other parameter values of importance that have no direct link to the SAM are the Harrod-neutral rate of technical change, which we denote x, and the labor force growth rate, n.

Our choice of the inter-temporal elasticity of substitution is based on the work of Giovannini (1985) and Kıpıcı (1996). Giovannini finds the inter-temporal elasticity of substitution to be strictly smaller than one which implies a value of $\theta > 1$. Kıpıcı (1996) estimated the value of θ for Turkey ranged between 1 and 1.27. The time preference rate ρ and depreciation δ are both set to a value of 0.04 (see Kydland and Prescott, 1982). The value of x is derived using growth accounting and data from the World Development Indicators (WDI), while n comes from either country demographic projections or using average labor force growth rates from the WDI. See Jorgenson (2005) for an introduction to the vast body of literature on growth accounting.

Each of the numerical SAMs introduced below originate from the 2001 Global Trade Analysis Project (GTAP) data base – with all data representing input-output and national account data in trillions of year 2001 Turkish lira. These data permit estimation of the production technologies used in each empirical model presented in Chapters 3, 4, 5, and 6, as well as the consumption shares used in the models. For example, the empirical SAM in Table 8.3 is used to identify the production parameters and consumption shares for the empirical model associated with the two-sector model of Chapter 3. The SAM in Section 8.5 has the information necessary to calculate the tax rates used to raise government revenue, as well as information necessary to calculate the composite capital function. The careful reader will see that each empirical SAM in this chapter is directly related to the empirical SAM of Section 8.5. For example, the SAM in Table 8.8 is a more aggregated version of the SAM in Table 8.11, while the SAM in Table 8.6 is a more aggregated version of that in Table 8.8.

This chapter is only intended to serve as a general introduction to SAMs, and to provide the reader with relatively detailed examples of how a SAM is linked to a corresponding empirical model. We do not discuss the question of how to construct a SAM. Nor do we discuss how to aggregate a large, multi-sector SAM into a three or four sector SAM that meets the modeler's needs. An adequate treatment of these issues is beyond the scope of this book. Still, the reader should find this chapter adequate for understanding the role of SAMs in dynamic, applied general equilibrium analysis.

The sections in this chapter follow closely the organization of the preceding chapters. The next section presents a SAM for the two-sector model of Chapter 3. We use this SAM to provide the reader with a relatively simple, non-technical description of the SAM and its organization, and then provide the reader with a precise definition of cell entries, and row – column sums. Section 8.3 presents a SAM for the basic three-sector model of Chapter 4, Section 8.4 presents a SAM for the three-sector model with intermediate inputs presented in Chapter 5, and Section 8.5 presents a SAM for the more policy oriented model of Chapter 6. Each section provides a numerical example using year 2001 national accounts data from Turkey. The last section summarizes the chapter discussions.

8.2 A two-sector, closed economy SAM

The model in Chapter 3 is a two-sector closed economy model, with two productive factors, labor and capital. In this stylized model, there is no government and, being a closed economy, there is no trade in final goods or capital accounts. A general SAM for such a model is presented in Table 8.1, which has five account categories: activity, commodity, factor, agent, and accumulation. Activity accounts summarize the economic costs (column entries) and receipts (row entries) for production activities. Commodity accounts summarize the value of market demand (column entries) and supply (row entries) of the goods and non-factor services produced in the economy (activities). Factor accounts summarize the payments to factors (column entries) and income received for the use of capital and labor services (row entries). The agent account gives the sources (row entries) and uses (column entries) of income, while the accumulation account gives the supply (row entries) and demand (column entries) of investment income.

Without additional data, the SAM provides no price information: all input and output prices are treated as being equal to unity. Later, we see that this feature of prices in SAMs changes slightly when tariffs and other policy distortions – and hence, government – are introduced into the model. Note that the organization of the SAM in Table 8.1 implies good 1 is a consumption good only, while good 2 is both consumed and used as investment capital, i.e., saved.

8.2.1 A non-technical description of the two-sector SAM

As noted above, when moving down a column, a column entry represents the value of expenditures paid from one account to another – when moving along a row, a row entry represents the value of receipts one account receives from another. Consider first, the interpretation of a typical cell column entry, where the intersection of an account column (e.g. Activity 1) with an account row (e.g., Labor) gives the value of payment made from the row account to the column account. The first account, "Activity," has two columns that identify the production activities 1 and 2. For the moment, imagine activity 1 represents the agricultural sector and activity 2 represents the rest of the economy. Each cell down the column "Activity 1" records the total payment from the agricultural sector to an account category. For example, the cell entry "Wages_1" is the *value* of wages paid to the flow of labor services used in sector 1, while the cell entry "Rent_1" is the *value* of rent paid for the flow of capital services used in sector 1. Thus, in this example, the column sum for Activity 1

$$\text{Cost}_1 = \text{Wages}_1 + \text{Rent}_1$$

Table 8.1 Structure of a two-sector Social Accounting Matrix with no trade

	Activity		Commodity		Factor		Agent	Accumulation	Total
	1	2	1	2	Labor	Capital	Household		
Activity									
1			$Sales_1$						$Production_1$
2				$Sales_2$					$Production_2$
Commodity									
1							$Cons_1$	Investment	$Demand_1$
2							$Cons_2$		$Demand_2$
Factor									
Labor	$Wages_1$	$Wages_2$							Lab Inc
Capital	$Rent_1$	$Rent_2$							Cap Inc
Agents									
Household					Wages	Rent			Income
Accumulation							Savings		Savings
Total	$Cost_1$	$Cost_2$	$Supply_1$	$Supply_2$	Lab Pay	Cap Pay	Expenditure	Investment	

is the total cost of labor and capital used to produce output 1; it is the value-added of sector 1 or the value of sector 1 output. A similar interpretation of cell entries extends to "Activity 2."

Next, consider the row entries corresponding to activity accounts. Again, there is a row associated with activity accounts 1 and 2. The intersection of Activity 1*s* row and Commodity 1'*s* column is the value of domestic demand of sector 1 output. Hence, "Sales_1" is the value of sector 1 demand, while "Sales_2" is the value of sector 2 demand. The double entry account system requires that total receipts be equal to total costs, and hence the row sum must be equal to the column sum, i.e.,

$$\text{Demand}_1 = \text{Supply}_1$$

Commodity accounts summarize the value of market transactions associated with the products generated in activities 1 and 2. With no trade, the column entry for commodity j is simply the value of domestic sales of good j, "Sales_j," where $j = 1, 2$. Here, moving down the column of Commodity 1, "Sales_1" is the market value of payments to producers of good 1. Of course, with constant returns to scale, in a competitive equilibrium and closed economy, "Sales_1" is exactly equal to "Cost_1," the cost of producing good 1. Moving along the row of Commodity j, the entry "Sales_j" is simply the total revenue received by producers of good j.

Consider next the two factor columns, labor and capital. The income from factors accrue to households. In the "Labor" column, the entry "Wages" is the value of wages paid to households in return for their labor services. In the "Capital" column, "Rent" is the value of rent paid to households for use of their capital stock. In the factor rows, "Wages_1" ("Wages_2") is the value of wages paid for labor services rendered to sector 1 (sector 2), while "Rent_1" ("Rent_2") is the value of rental payments for capital services rendered to sector 1 (sector 2). As with the activity and commodity accounts, the SAM's double entry nature requires that

$$\text{Lab Pay} = \text{Lab Inc}$$

i.e., the value of labor payments from firms to households is exactly equal to the value of income households receive for their labor services. Similar arguments hold for capital rental payments and rental income.

Moving down the Household column, we have total expenditures on consumption goods 1 and 2, and the non-consumption income diverted to savings. Here, "Cons_j" is the domestic household expenditure on good $j = 1, 2$. "Savings" is the amount of its income the household makes available (payment) to the capital market. Notice the savings of domestic households appears in a separate row referred to as "Accumulation." The household column entries shows that income is spent on consumption goods and savings. Moving along the Household row we find the total factor payments to households for the use of their labor and capital services. Here, "Wages" is the total value of wages received by households in exchange for labor service flows, while "Rent" is the total value of capital rent received by households in exchange for capital services flows.

The SAM alone does not specify or depict any behavioral and institutional characteristics of a market economy. Instead, the SAM is a collection of identities over economic objects. For example, if it is not yet obvious,

$$\text{Cost}_1 + \text{Cost}_2 = \text{Demand}_1 + \text{ Demand}_2 = \text{ Lab Inc} + \text{Capital Inc}$$

where the first sum is valued added GDP, the second sum is the expenditure measure of GDP, and the last sum is the income measure of GDP. This income is used for consumption or savings. In other words, the SAM organizes data based on economic identities which, in turn, we link to a model that exploits the myriad of relationships among economic variables.

8.2.2 A more technical description of the two-sector SAM

We now turn to a more technical discussion of the SAM. Included in this discussion is the role of budget constraints and market clearing. Here we give a precise definition of prices and cell entries, and show the conditions that ensure the sum of row account entries are equal to the sum of corresponding column account entries.

The first and the most important assumption typically imposed on the data is that it was generated by a market clearing–equilibrium process. If the SAM is constructed for a static general equilibrium analysis, the data reflected in the SAM are typically interpreted as the outcome of an economy in equilibrium. The existence of an equilibrium not only implies the data organized in the SAM are balanced, but also that they are derived from the outcomes satisfying the rationality assumption of agents in the economy, i.e., consumers maximize utility and firms maximize profit. This important assumption allows us to calibrate (i.e., estimate) many of the key parameters of the model in such a way that when we solve the resulting model, it will reproduce the data exactly. This property also becomes an important diagnostic tool for uncovering modeling and coding errors.

In Table 8.2, the following two market clearing conditions are exploited

$$Y_1 = Q_1 + \dot{K} \quad (1)$$

$$Y_2 = Q_2 \quad (2)$$

Here, Q_1 is aggregate quantity of good 1 demanded by the household, Q_2 is aggregate quantity of good 2 demand by the household, and $\dot{K}$ is the quantity of good 1 invested. The value of total investment is equal to $p_1\dot{K}$, while the value of total consumption is equal to $p_1Q_1 + p_2Q_2$. Throughout this chapter, technologies are assumed to satisfy constant returns to scale. It follows that the total value of production received by producers is equal to the payments to all factors of production. Also, one advantage of these units is we can set each output price p_j equal to unity in the base case equilibrium. As Kehoe (1996) suggests, we can think of these variables as price indices, which are naturally set equal to one in the base period. We can also index the inital labor supply to unity so that w is total wage payments. Later, we also index the agricultural land endowment to

Table 8.2 The two-sector SAM with more detail

	Activity		Commodity		Factor		Agents	Accumulation	Total
	1	2	1	2	Labor	Capital	Household		
Activity									
1			Y_1						pY_2
2				pY_2					pY_2
Commodity									
1							Q_1	$\dot{K}$	$Q_1 + \dot{K}$
2							pQ_2		pQ_2
Factor									
Labor	wL_1	wL_2							wL
Capital	$r^k K_1$	$r^k K_2$							$r^k K$
Agents									
Household					wL	$r^k K$			$G(L, K)$
Accumulation							$\dot{K}$		$\dot{K}$
Total	$C^1(\cdot) Y_1$	$C^2(\cdot) Y_2$	Y_1	pY_2	wL	$r^k K$	$\mathcal{E}(p) Q + \dot{K}$	$\dot{K}$	

unity and then treat land rental payments π as total payments to land.

The sum of all Activity j column entries is $C^j(\cdot)Y_j = wL_j + r^kK_j$, the total cost of producing Y_j units of good j. The sum of all Activity j row entries is p_jY_j. With constant returns to scale, it follows that $C^j(\cdot)Y_j = p_jY_j$. In other words, the market value of sector j output is exactly equal to the cost of producing that output. The sum of all Commodity j column entries is simply p_jY_j, the market value of good j supply. The sum of all Commodity j rows is equal to the receipts from the sale of commodity j, i.e., the value of consumption, or in the case of commodity 1, the value of consumption plus investment.

The values wL_j and r^kK_j are the rents sector j paid to capital and labor, respectively, and $L = L_1 + L_2$,while $K = K_1 + K_2$. As in Chapter 3, r^k and w represent the rate of return to capital and the wage rate respectively. The sum of all cells in the Household row account is total income, $GDP = wL + r^kK$. The sum of all Household column accounts is expenditure plus savings, $\mathcal{E}(p_1, p_2)Q + \dot{K}$, where $\dot{K}$ is savings, and $\mathcal{E}(\cdot)Q$ is household expenditures on goods 1 and 2. Of course, in equilibrium the following conditions are satisfied:

$$\begin{aligned} G(L,K) &= \underbrace{\mathcal{E}(p)Q + \dot{K}}_{\text{Expenditures}} = \underbrace{wL + r^kK}_{\text{Income}} \\ &= \underbrace{C^1(\cdot)Y_1 + C^2(\cdot)Y_2}_{\text{Value Added}} = Y_1 + pY_2 \end{aligned}$$

where the above equalities suggest the equivalence of the expenditure, income, and value-added approaches to measuring GDP.

8.2.3 Using the SAM to calibrate the empirical two-sector model

This section illustrates how to parameterize the empirical analog of the theoretical model discussed in Chapter 3, Section 4.2. The major components of the empirical model are its primitives – the production technologies and household preferences. Of course, the production technologies provide the foundations for determining output supply and input demands, while household preferences determine the final good demand of the economy. The parameters of the technologies and underlying preferences are based upon the social accounting matrix of Table 8.3.

As noted above, when using the SAM to calibrate a static model, the SAM data are typically viewed as a one period snapshot of an economy in equilibrium. A static model is calibrated to these data in such a way that a base-run solution of the numerical model exactly reproduces the SAM. Dynamic models require additional information that characterize inter-temporal equilibrium – information that includes behavioral parameters not present in the SAM. Hence, it is quite possible the solution to a dynamic model will not exactly reproduce the base period data. For the purposes of this discussion, base period data is viewed as an equilibrium point on a country's transition path to long-run equilibrium.

Labor in the SAM

Recall that in Chapter 3, the total amount of labor is normalized to unity, i.e., $l_j = L_j/L$, with $l_1 + l_2 = l = 1$. Since $l_1 + l_2 = 1$ in equilibrium, it is natural to interpret the entries along the "Labor" row account as the value of wages received by labor in Activities 1 and 2. In other words, the value 80268.4 is equal to $w \cdot l = w$, implying $w = 80268.4$ represents total household wage income. It naturally follows that households received 27535.1 for labor services rendered to sector 1 and 52733.4 for labor services rendered to sector 2. Equally important is that the "amount" of labor demanded by sector j is simply the share of labor income paid by sector j, i.e., $l_1 = 27535.1/80268.4 = 0.34304$ and $l_2 = 1 - l_1 = 0.65696$. After obtaining a numerical solution, the total wage bill w, is easily converted back to wage per worker per year based upon country employment level data.

Estimating the capital stock

The discussion above describes how to derive a reasonable estimate of the stock of labor used by each sector. Normalizing the

Table 8.3 A two-sector SAM for Turkey (trillion of 2001 Turkish lira)

	Activity		Commodity		Factor		Agents	Accumulation	Total
	1	2	1	2	Capital	Labor	Household		
Activity									
1			57207.1						57207.1
2				99629.1					99629.1
Commodity									
1							52511.6	4695.5	57207.1
2							99629.1		99629.1
Factor									
Capital	29672.0	46895.7							76567.7
Labor	27535.1	52733.4							80268.4
Agents									
Household					76567.7	80268.4			156836.2
Accumulation							4695.5		4695.5
Total	57207.1	99629.1	57207.1	99629.1	76567.7	80268.4	156836.2	4695.5	

stock of labor to unity simplified things considerably, as it enabled us to interpret each entry along the "Labor" row account as a total wage, or a fraction of that total wage; e.g., 80268.4 is the total wage paid to labor, while "Activity 1" entry for that row, 27535.1, is 34.3 percent of the total wage. Developing an estimate of the stock of capital, however, is more challenging because each entry along the "Capital" row account represents the capital rental payments received by households for the flow of capital services they provided, where the value of these rental payments is the product of a rate of return to capital and a capital stock level. The implied warning here is to not confuse capital rental payments with capital stocks, and remember that production technologies are defined over the levels of labor and physical capital employed, not over the value of payments to these factors.

Given each entry along the capital row account represents capital rental receipts, the question remains how to estimate an economy's capital stock level. There are several approaches to capital stock estimation. Although a thorough treatment of this topic is beyond the scope of this chapter,[2] below we provide a short discussion of three approaches. We forewarn the reader, however, that estimating a capital stock can prove challenging.

Perhaps the easiest approach to choosing a country's capital stock level, is to use estimates from existing government sources or estimates from reputable studies. For instance, the United States Bureau of Economic Analysis (BEA) publishes a time series on fixed assets.[3] Assuming well-functioning capital markets, the implied assumption is each sector pays the same rate of return on capital. Under such an assumption, the allocation of capital stock across sectors is proportional to the share of rental receipts across those sectors. For example, in Table 8.3, the total value of asset rent is Lira 76567.7 billion. Note, the value 76567.7 includes the value of rental payments to both land and physical

---

[2]An early references to capital stock measurement is Jorgenson and Griliches (1967), and Usher (1980). See also Jorgenson (2005, pp. 759–761).

[3]See the BEA's National Economic Accounts: http://www.bea.gov/national/Index.htm

assets (capital stock), with the value of land rental payments equal to Lira 1975.6 billion. Sector 1's share of rental payments is equal to 29672.0/76567.7 = 0.38753, while sector 2s share of rental payments is equal to 1 – 0.38753 = 0.61247. Hence, if the published capital stock level for Turkey in 2001 is Lira 621 trillion, then the estimate of capital stock employed by sector j is the product of the sector's share and this stock.

In the event capital stock or fixed asset data are not available, a second approach is to use the perpetual inventory method (see Hall and Jones, 1999) to estimate the capital stock. The perpetual inventory method uses the following relationship:

$$K_{t+1} = K_t - \delta K_{t+1} + I_{t+1} \tag{8.1}$$

where K_t and I_t are time t capital stock and investment, and δ is depreciation. In practice, the variable I_t is typically approximated by gross fixed capital formation in local currency units (LCU), a time series available for most countries in the World Bank's WDI. The parameter δ is either estimated by the analyst or chosen from a reputable study. Equation (8.1) is a recursive expression that, in addition to I, requires an initial level of K to solve. Hall and Jones (1999) suggest estimating the initial capital stock level using:

$$K_0 = \frac{I_0}{g + \delta}$$

where g is the country's average rate of growth in gross fixed capital formation or its rate of growth in GDP. One then uses an appropriate gross fixed capital series to estimate a corresponding aggregate capital stock level for the SAM. Using the perpetual inventory method, our estimate of the capital stock in 2001 is K_{2001} = Lira 621, 938 trillion, while the estimate of the stock of assets (in capital stock equivalents) is equal to Lira 638418 billion.[4]

[4] With K_{2001} = 621938.04 and capital rental payments equal to 74591.2, it follows that an estimate of r^k_{2001} = 74591.2/621938.04 = 0.119933. With land normalized to unity, we convert land into a capital stock equivalent by dividing the land rental payment by the rate of return to the asset capital: 1976.5/0.119933 = 16480. Adding this value to 621938 yields the asset stock estimate.

A third approach to estimating the capital stock is to calculate a value for the underlying rate of return to capital, and then derive an estimate of the stock of capital using the relationship $K_{2001} = r^k K / r^k_{2001}$. Here K_{2001} is the estimate of the stock of capital for 2001, $r^k K$ is the total value of capital rents as given by the SAM, and r^k_{2001} is the estimate of the initial rate of return to capital. Similarly, the stock of capital employed by sectors 1 and 2 are estimated by $r^k K_1 / r^k_{2001}$ and $r^k K_2 / r^k_{2001}$ respectively. One way to calculate the initial rate of return to capital is to use household consumption data on the rate of growth in consumption and exploit the Euler equation

$$\frac{\dot{\epsilon}}{\epsilon} = \frac{1}{\theta}\left[r^k - \delta - \rho - (1-\lambda)(1-\theta)\frac{\dot{p}}{p}\right]$$

which implies

$$r^k = \theta\frac{\dot{\epsilon}}{\epsilon} + \delta + \rho + (1-\lambda)(1-\theta)\frac{\dot{p}}{p}$$

Using WDI data we calculated $\dot{\epsilon}/\epsilon$ using the growth of household final good consumption expenditure per worker: between 2000 and 2001, $\dot{\epsilon}/\epsilon = 0.0261$. Given p is the relative price of good 2 with respect to good 1, we assume – over a one year period – the change in relative prices is small enough to ignore, i.e., $\dot{p}/p = 0$. Using a generalized method of moment estimation procedure, Kıpıcı (1996) estimates the value of θ for Turkey falls between 0.99 and 1.27. In the empirical model we assume $\theta = 1.26$ and set $\rho = 0.04$ and $\delta = 0.04$ (see Kydland and Prescott, 1982). Using $\dot{\epsilon}/\epsilon = 0.0261$, $\dot{p}/p = 0$, and the parameter values $\theta = 1.26, \rho = \delta = 0.04$, and $\lambda = 0.3451$, our estimate of the rate of return to capital in 2001 would be $r^k_{2001} = 0.1129$. This value for the rate of return to capital yields capital stock estimates equal to

$$K_{2001} = \frac{r^k K}{r^k_{2001}} = \frac{74591.2}{0.1129} = 660684$$

In calibrating the two-sector Ramsey model, we use the stock of assets in capital equivalents value of 638418 as our estimate of K. In the remaining empirical exercises the land and capital assets are disaggregated and, hence, use the capital stock level $621,938.04$ as our measure of K.

Model calibration

In each model presented in this book, we represent sector production technologies and household preferences by the Cobb-Douglas function. This functional form is chosen primarily for its ease of exposition. The Cobb-Douglas technologies for sectors 1 and 2 are:

$$\begin{aligned} Y_1 &= \Psi_1 (K_1)^{1-\alpha} (l_1)^{\alpha} \\ Y_2 &= \Psi_2 (K_2)^{1-\beta} (l_2)^{\beta} \end{aligned} \tag{8.2}$$

As noted in Chapter 2, with constant returns to scale technologies, the production elasticity α is exactly equal to the cost share of labor in producing good 1. It follows that our estimated value of α, denoted $\hat{\alpha}$, is given by

$$\hat{\alpha} = \frac{wL_1}{C^1 (w, r^k) Y_1} = \frac{27535.1}{57207.1} = 0.4813$$

Similarly, $\hat{\beta}$, the estimate of labor's share of cost in producing good 2 is given by

$$\hat{\beta} = \frac{wL_2}{C^2 (w, r^k) Y_2} = \frac{52733.4}{99629.1} = 0.5293$$

Finally, using $r^k_{2001} = 0.1199$, derive the scaling parameter $\hat{\Psi}_1$ by solving the following problem:

$$\hat{\Psi}_1 = \frac{Y_1}{(l_1)^{\hat{\alpha}} \left(\frac{r^k K_1}{r^k_{2001}}\right)^{1-\hat{\alpha}}} = \frac{57207.1}{0.343030^{0.4813} \left(\frac{29672}{0.11993}\right)^{1-0.4813}} = 152.59$$

and

$$\hat{\Psi}_2 = \frac{Y_2}{(l_2)^{\hat{\beta}} \left(\frac{r^k K_2}{r^k_{2001}}\right)^{1-\hat{\beta}}} = \frac{99629.1}{0.65696^{0.5293} \left(\frac{46895.7}{0.11993}\right)^{1-0.5293}} = 290.21$$

It follows that the sectoral technologies are given by

$$Y_1 = 152.59\,(l_1)^{0.4813}\,(K_1)^{0.5187} \tag{8.3}$$
$$Y_2 = 290.21\,(l_2)^{0.5293}\,(K_2)^{0.4707} \tag{8.4}$$

The cost functions corresponding to expressions (8.3) and (8.4) are[5]

$$C^1\left(w, r^k\right) Y_1 = 0.0130937 w^{0.4813}\left(r^k\right)^{0.5187} Y_1 \tag{8.5}$$
$$C^2\left(w, r^k\right) Y_2 = 0.0068799 w^{0.5293}\left(r^k\right)^{0.4707} Y_2 \tag{8.6}$$

To calibrate the felicity function, the account column of interest is "Agent." Using the homothetic Cobb-Douglas representation of felicity, $u = (Q_1)^{\lambda}(Q_2)^{1-\lambda}$, where Q_i is aggregate consumption of good $i = 1, 2$. Here, the parameter λ is simply the share of consumption income spent on good 1. From the discussion directly above

$$\hat{\lambda} = \frac{pQ_1}{pQ_1 + Q_2} = \frac{52511.6}{52511.6 + 99629.1} = 0.3452$$

and the parameterization of the corresponding felicity function is

$$u = (Q_1)^{0.3452}\,(Q_2)^{0.6548} \tag{8.7}$$

As an exercise, the reader can verify that given cost and felicity functions (8.5), (8.6), and (8.7), if the model economy is endowed with 68.5 million units of labor, 638,412 (trillion) units of capital, and households save Lira trillion 4695.5, then in a competitive equilibrium, the model will generate the data observed in Table 8.3.

---

[5] A useful check on software script (e.g., Mathematica coding) is to substitute the market clearing wage rate and rate of return to capital into Equations (8.5) and (8.6), and verify the following zero profit conditions are satisfied

$$0.0130937W\,(1)^{0.4813}\,R\,(1)^{0.5187} = 1$$
$$0.0068799W\,(1)^{0.5293}\,W\,(1)^{0.4707} = 1$$

We suggest making use of the many empirical functions – like Equations (8.3) and (8.4) – and equilibrium conditions of your model to check for scripting errors.

8.3 A three-sector, open economy SAM

Chapter 4 presents a three-sector, open economy model. The economy employs three factors to produce three outputs. Here, the factors are labor, capital, and the sector specific factor, land, while the outputs are agriculture, manufacturing, and a home-good that in the empirical example we refer to as services. Since the model in Chapter 4 is that of an open economy, it allows for the possibility of trade in final goods. As in the two-sector model, here government consumption is aggregated into household consumption.

Table 8.4 presents the SAM with no intermediate factors, and includes six account categories: activities, commodities, factors, agents, trade, and accumulation. Table 8.4 adds to Table 8.3: (i) an activity and commodity sub-account for services, (ii) a factor sub-account for land rent, and (iii) a trade account. Also, with the introduction of trade, we distinguish between the domestic and foreign production (imports) of a traded good, and domestic and foreign consumption (exports) of that good.

The addition of services, land, and trade introduces several new cell entries. The discussion below moves down a column account and highlights the new column cell entries for that account. We begin by first moving down the Activity a column, then move rightward across the SAM column accounts until we reach the Total Receipts column. We leave it to the reader to interpret the row column entries.

The only new entry in the column account "Activity a" is "Land Rent," which is the rental income agriculture pays households for the use of land services. Moving over to the column account, "Activity s," the cell entries "wage_s" and "rent_s" are labor and capital rental costs incurred by the service sector in producing its output. The cell entry "Cost_s" is simply the total cost of producing the service good.

With the introduction of trade, we must now distinguish between the domestic and foreign consumption and production of goods. Moving down the column account, Commodity-j, the cell entry "DomSale_j" is the smaller of the value of commodity-

Table 8.4 Structure of a three-sector social accounting matrix with trade

	Activity			Commodity			Factors			Agents	Accumulation	Trade	Total Receipts
	m	a	s	m	a	s	Capital	Labor	Land	Household			
Activity:													
Manufacture (m)				DomSale_m								Export_m	$p_m Y_m$
Agriculture (a)					DomSale_a							Export_a	$p_a Y_a$
Services (s)						Sales_s							Sales_s
Commodity:													
Manufacture (m)										DomCons_m	$\dot{K}$		Absorp_m
Agriculture (a)										DomCons_a			Absorp_a
Services (s)										Con_s			Absorp_s
Inputs													
Capital	$r^k K_m$	$r^k K_a$	rent_s										$r^k K$
Labor	$w L_m$	$w L_a$	wage_s										$w L$
Land		LandRent											LandRent
Agents													
Household							$r^k K$	$w L$	LandRent				$G(\cdot)$
Accumulation										$\dot{K}$			$\dot{K}$
Trade				Import_m	Import_a								ForeignExp
Total Expenditure	$C^m(\cdot) Y_m$	$C^a(\cdot) Y_a$	Cost_s	Absorp_m	Absorp_a	Sales_s	$r^k K$	$w L$	LandRent	$E(\cdot) Q + \dot{K}$	$\dot{K}$	ForeignEarn	

j both produced and consumed within the region. The column sum of commodity-j is often referred to as the domestic absorption of good j. The intersection of the Total Expenditure row with the Commodity a and m accounts is typically referred to as absorption, and is the value of domestic plus foreign sales of each respective good in the economy. For example, "Absorp$_a$" is the sum of the value of agriculture produced and sold in the region plus the value of agriculture imported and sold in the region. A more precise definition of domestic absorption follows shortly.

The cell "Import$_j$" is the market value of sector j imports, i.e., the value of commodity-j produced abroad, but consumed domestically. "Absorb$_i$" is the total value of good j consumption, $j = a, m$. The cell entry "Sales$_s$" is the market value of non-traded good consumption (and production). The new factor account "Land" has the cell entry "LandRent," which is simply the total value of rent paid to landowners.

Moving to the "Agent Household" column account, observe that the cell entries for agriculture and manufacturing are renamed. For example, the new name "DomCons$_a$" stresses that the cell now measures total household expenditure on domestic agricultural consumption: a value that will typically differ from domestic production values. The cell entry "Cons$_s$" is the total household expenditure on non-traded good consumption.

The new account "Trade" has cell entries for activity accounts a and m. "Export$_a$" is the rest of the world's expenditure on the region's or country's agricultural output, while "Export$_m$" is the rest of the world's expenditure on the region's manufacturing output. The total rest-of-the-world expenditure and the region's output is "ForeignEarn." This SAM implicitly assumes there is no trade in services. Note that in this model, there is no foreign ownership of capital, as indicated by a lack of entries in the cell entries where the "Trade" and "Accumulation" accounts intersect.

Using notation consistent with Chapter 4, Table 8.5 presents the three-sector SAM with no intermediate factors corresponding to Table 8.4. Here, πH is total land rents, where π is the per unit land rent and H is the region's land endowment, which is normalized to unity. The variables IM_i and EX_j are the level of imports and exports of good $j = a, m$, while Q_j^{DA} is the domestic absorption of good $j = a, m$, where

$$\begin{aligned} p_a Q_a^{DA} &= \min \{p_a Q_a, p_a Y_a\} \\ Q_m^{DA} &= \min \left\{ Q_a + \dot{K}, Y_a \right\} \end{aligned}$$

and where each output price is normalized by the manufacturing price p_m. Finally, $Y_j^{AB} = p_j \left(Q_j^{DA} + IM_j\right)$ is the absorption of good $j = a, m$. The remaining variables are as defined in Chapter 4, e.g., L_s is service sector labor demand.

As in the two-sector example in the prior section, several equilibrium conditions hold in Table 8.5. First, for activity accounts, the constant returns to scale assumption requires

$$\begin{aligned} C^j \left(w, r^k\right) Y_j &= p_i Y_i \qquad i = m, s \\ C^a \left(w, r^k, \pi\right) Y_a &= p_a Y_a \end{aligned}$$

i.e., in equilibrium the total cost of producing good j is equal to the revenue generated by the sale of good j. Second, for commodity accounts the market value of domestic demand for each good is equal to the market value of domestic supply of each good, i.e.,

$$\begin{aligned} p_j Q_j^{DA} &= p_j Y_j^{AB} \qquad j = a, m \\ p_s Q_s &= p_s Y_s \end{aligned}$$

The above conditions are, of course, domestic market clearing conditions. Third, aggregate household income (GDP) is equal to the total consumption expenditures plus savings: $wL + r^k K + \pi H = \mathcal{E}\left(p_a, p_m, p_s\right) q + \dot{K}$. We also have the value of exports is equal to the value of imports:

$$\sum_{j=a,m} p_j EX_j = \sum_{j=a,m} p_j IM_j$$

Finally, the value of payments for labor (capital, land) services from each sector is equal to the household income received for labor (capital, land) services.

Table 8.5 The three-sector SAM with trade, with more detail

	Activity			Commodity			Factors			Agents	Accumulation	Trade	Total Receipts
	m	a	s	m	a	s	Capital	Labor	Land	Household			
Activity:													
Manufacture (m)				Q_m^{DA}								EX_m	Y_m
Agriculture (a)					$p_a Q_a^{DA}$							$p_a EX_a$	$p_a Y_a$
Services (s)						$p_s Y_s$							$p_s Y_s$
Commodity:													
Manufacture (m)										Q_m	$\dot{K}$		Q_m^{DA}
Agriculture (a)										$p_a Q_a^{DA}$			$p_a Q_a^{DA}$
Services (s)										$p_s Q_s$			$p_s Q_s$
Inputs													
Capital	$r^k K_m$	$r^k K_a$	$r^k K_s$										$r^k K$
Labor	wL_m	wL_a	wL_s										wL
Land		πH											πH
Agents													
Household							$r^k K$	wL	πH				$G(L, K, H)$
Accumulation										$\dot{K}$			$\dot{K}$
Trade				IM_m	$p_a IM_a$								$\sum_{i=a,m} IM_i$
Total Expenditure	$C^m(\cdot) Y_m$	$C^a(\cdot) Y_a$	$C^s(\cdot) Y_s$	Y_m^{AB}	$p_a Y_a^{AB}$	$p_s Y_s$	$r^k K$	wL	πH	$\mathcal{E}(\cdot) Q + \dot{K}$	$\dot{K}$	$\sum_{j=a,m} EX_j$	

8.3.1 Using the SAM to calibrate the empirical three-sector model

Table 8.6 presents the SAM underlying the empirical three-sector model in Chapter 4. Note, in Table 8.6 "Household," "Accumulation," and "Total Expenditures" are abbreviated with "HH," "Accum," and "Tot Exp" respectively.

Introducing trade has no effect on the strategies for estimating the parameter values of the preference and production technologies. Consider first, the case of homothetic preferences. With a Cobb-Douglas felicity function, preferences are given by $u(Q_a, Q_m, Q_s) = Q_a^{\lambda_a} Q_m^{\lambda_a} Q_s^{1-\lambda_a-\lambda_m}$. Here, Q_i is the aggregate consumption of good $j = a, m, s$, and λ_j is the consumption share of good j. By Table 8.6, the consumption share estimates are $\hat{\lambda}_a = 0.18203$ and $\hat{\lambda}_m = 0.16312$, giving felicity function

$$u(Q_a, Q_m, Q_s) = Q_a^{0.182} Q_m^{0.163} Q_s^{0.655} \tag{8.8}$$

With $p_m = 1$ as numeraire, the corresponding expenditure function for (8.8) is

$$\mathcal{E}(p_a, p_s) Q \equiv \min_{Q_a, Q_m, Q_s} \left\{ p_a Q_a + Q_m + p_s Q_s : u \leq Q_a^{0.182} Q_m^{0.163} Q_s^{0.655} \right\}$$

the derivation of which is left to the reader.

The Cobb-Douglas technologies for sectors m and s are:

$$\begin{aligned} Y_m &= \Psi_m (l_m)^{\alpha} (K_m)^{1-\alpha} \\ Y_s &= \Psi_s (l_s)^{\beta} (K_s)^{1-\beta} \end{aligned}$$

It follows that our estimated value of α, denoted $\hat{\alpha}$, is given by

$$\hat{\alpha} = \frac{wL_m}{C^m(w, r^k) Y_m} = \frac{14129.7}{32403.3} = 0.4361 \tag{8.9}$$

Similarly, $\hat{\beta}$, the estimate of labor's share of cost in producing services is given by

$$\hat{\beta} = \frac{wL_s}{C^s(w, r^k) Y_s} = \frac{52733.4}{99629.1} = 0.5293 \tag{8.10}$$

Table 8.6 A three-sector SAM for Turkey (trillion 2001Turkish Lira)

	Activity			Commodity			Factors			Agents	Accum	Trade	Tot Receipts
	m	a	s	m	a	s	K	L	H	HH			
Activity:													
m				29512.4								2890.9	32403.3
a					24803.8								24803.8
s						99629.1							99629.1
Commodity:													
m										24816.9	4695.5		29512.4
a										27694.7			27694.7
s										99629.1			99629.1
Inputs													
K	18273.5	9422.0	46895.7										74591.2
L	14129.7	13405.3	52733.4										80268.4
H	0.0	1976.5	0.0										1976.5
Agents													
HH	0.0	0.0	0.0				74591.2	80268.4	1976.5				156836.2
Accum										4695.5			4695.5
Trade				0.0	2890.9								2890.9
Tot Exp	32403.3	24803.8	99629.1	29512.4	27694.7	99629.1	74591.2	80268.4	1976.5	156836.2	4695.5	2890.9	

With $L = 1$ and $l_j = L_j/L$, straightforward calculations yield $l_m = 0.1760$, $l_s = 0.65696$, and $l_a = 0.1670$. Using $r^k_{2001} = 0.11993$, it follows that the sectoral technologies are given by

$$\begin{aligned} Y_m &= 82.5496\,(l_m)^{0.4361}\,(K_m)^{0.5639} \\ Y_s &= 290.201\,(l_s)^{0.5293}\,(K_s)^{0.4707} \end{aligned}$$

and the corresponding cost functions are

$$\begin{aligned} C^m\left(w, r^k\right)Y_m &= 0.02403w^{0.4361}\left(r^k\right)^{0.5639}Y_m \\ C^s\left(w, r^k\right)Y_s &= 0.00687995w^{0.5293}\left(r^k\right)^{0.4707}Y_s \end{aligned}$$

Consider the agricultural technology

$$Y_a = \Psi_a\,(l_a)^{\phi_1}\,(K_a)^{\phi_2}\,H^{1-\phi_1-\phi_2} \tag{8.11}$$

Then the agricultural factor cost share estimates are:

$$\begin{aligned} \hat{\phi}_1 &= \frac{wL_a}{C^a(\cdot)\,Y_a} = \frac{13405.3}{24803.8} = 0.54045 \\ \hat{\phi}_2 &= \frac{r^kK_a}{C^a(\cdot)\,Y_a} = \frac{9422.0}{24803.8} = 0.37986 \\ &1 - \hat{\phi}_1 - \hat{\phi}_2 = 0.079769 \end{aligned}$$

Normalizing the land endowment equal to unity, $H = 1$, the corresponding scaling parameter is

$$\hat{\Psi}_a = \frac{24803.8}{\left(\frac{9422.0}{0.11993}\right)^{0.37986}\left(\frac{13405.3}{80268.4}\right)^{0.54045}} = 901.787$$

and the empirical analog of the agricultural technology is

$$Y_a = 901.787\,(l_a)^{0.5404}\,(K_a)^{0.3799}$$

The interested reader can show that with $H = p_a = 1$, the agricultural value-added – i.e., land rental function – corresponding to the technology (8.11) is

$$\begin{aligned} \pi^a(p_a, w, r^k)H &= \phi_1^{\frac{\phi_1}{\phi_3}}\phi_2^{\frac{\phi_2}{\phi_3}}\phi_3\,(p_a\Psi_a)^{\frac{1}{\phi_3}}\,w^{\frac{-\phi_1}{\phi_3}}\left(r^k\right)^{\frac{-\phi_2}{\phi_3}}H \\ &= 1.47795\times 10^{32}w^{-6.78234}\left(r^k\right)^{-4.767} \end{aligned}$$

8.4 A three-sector, open economy SAM with intermediate products

In Table 8.7, "Input$_{ij}$" $\equiv p_i Y_{ij}$, where $p_i Y_{ij}$ is the value of the quantity of good i employed as an intermediate input in producing good j. The sum $r^k K_m + wL_m$ is the value-added in manufacturing. Similarly, $r^k K_a + wL_a + \pi H$ and $r^k K_s + wL_s$ are value-added for the agricultural and service sectors. The column total TC_j is the total cost of producing good j, and is the sum of value-added costs and intermediate input costs. In the case where intermediate inputs enter in a Leontief fashion, "Input$_{ij}$" $= p_i Y_{ij} = p_i \sigma_{ij} Y_j$, where σ_{ij} is the input-output ratio of the amount of good i used to produce a unit of good j. For example, σ_{ma} is the number of units of good m required to produce a unit of agricultural output. Total cost is given by

$$TC_j = C^j\left(w, r^k\right) Y_j + p_a \sigma_{aj} Y_j + p_m \sigma_{mj} Y_j + p_s \sigma_{sj} Y_j, \quad j = a, m, s$$

where Y_j is now gross output.

8.4.1 Using the SAM to calibrate the empirical three-sector model with intermediate inputs

Table 8.8 presents the SAM underlying the empirical three-sector model with intermediate inputs in Chapter 5. The consumption levels here are identical to those in Table 8.6, hence the felicity function associated with Table 8.8 is identical to Equation (8.8). Introducing intermediate products changes, slightly, the production technology parameters.

The manufacturing and service sector's value-added technology is approximated by the Cobb-Douglas technology

$$Y_j = \Psi_j \left(l_j\right)^{\alpha_j} \left(K_j\right)^{1-\alpha_j}, \quad j = m, s$$

For $j = m$ the corresponding minimum production cost is

$$\left[C^m\left(w, r^k\right) + \sum_{i=a,m,s} \sigma_{im} p_i\right] Y_m =$$

Table 8.7 Structure of a three-sector social accounting matrix with trade and intermediate inputs

	Activity			Commodity			Factors			Agents	Accum	Trade	Tot Receipts
	m	a	s	m	a	s	K	L	H	HH			
Activity:													
m				Q_m^{DA}								EX_m	Y_m
a					$p_a Q_a^{DA}$							$p_a EX_a$	$p_a Y_a$
s						$p_s Y_s$							$p_s Y_s$
Commodity:													
m	Input_{mm}	Input_{ma}	Input_{ms}							Q_m	$\dot{K}$		Q_m^{DA}
a	Input_{am}	Input_{aa}	Input_{as}							$p_a Q_a^{DA}$			$p_a Q_a^{DA}$
s	Input_{sm}	Input_{sa}	Input_{ss}							$p_s Q_s$			$p_s Q_s$
Inputs													
K	$r^k K_m$	$r^k K_a$	$r^k K_s$										$r^k K$
L	wL_m	wL_a	wL_s										wL
H		πH											πH
Agents													
HH							$r^k K$	wL	πH				$G(L, K, H)$
Accum										$\dot{K}$			$\dot{K}$
Trade				IM_m	$p_a IM_a$								$\sum_{=a,m} IM_j$
Tot Exp	TC_m	TC_a	TC_s	Y_m^{AB}	$p_a Y_a^{AB}$	$p_s Y_s$	$r^k K$	wL	πH	$\mathcal{E}(\cdot) Q + \dot{K}$	$\dot{K}$	$\sum_{j=a,m} EX_j$	

Table 8.8 A three-sector SAM with intermediate inputs for Turkey (trillion Turkish Lira)

	Activity			Commodity			Factors			Agents	Accum	Trade	Tot Receipts
	m	a	s	m	a	s	K	L	H	HH			
Activity:													
m				97153.9									97153.9
a					42595.9							5222.9	47818.9
s						155893.9							155893.9
Commodity:													
m	46037.2	5385.3	23255.8							24816.9	2881.7		102376.8
a	1433.5	11141.2	2326.6							27694.7			42595.9
s	17279.9	6488.6	30682.4							101442.9			155893.9
Inputs													
K	18273.5	9422.0	46895.7										74591.2
L	14129.7	13405.3	52733.4										80268.4
H		1976.5											1976.5
Agents													
HH							74591.2	80268.4	1976.5				156836.2
Accum										2881.7			2881.7
Trade				5222.9									5222.9
Tot Exp	97153.9	47818.9	155893.9	102376.8	42595.9	155893.9	74591.2	80268.4	1976.5	156836.2	2881.7	5222.9	

$$\overbrace{\alpha_m^{-\alpha m}\left(1-\alpha\right)^{-(1-\alpha)} w^{\alpha}\left(r^k\right)^{1-\alpha}\frac{Y_m}{\Psi_m}}^{=C^m\left(w,r^k\right)Y_m} + \sum_{i=a,m,s} p_i\sigma_{im}Y_m \quad (8.12)$$

A similar expression obtains for the service sector. To develop the empirical analog of expression (8.12) requires estimating five parameters each, for sectors $j = m, s$: they are $\alpha, \Psi_m, \sigma_{am}, \sigma_{mm}$, and σ_{sm} for manufacturing and $\beta, \Psi_s, \sigma_{as}, \sigma_{ms}$, and σ_{ss} for the service sector. Estimates of the input-output coefficients σ_{ij} are derived using the following identity

$$\hat{\sigma}_{ij} \equiv \frac{p_i\sigma_{ij}Y_j}{TC_j}, \qquad j = m, s; \quad i = a, m, s$$

From Table 8.8 we get the following values

$$\begin{aligned}\sigma_{mm} &= 0.47359, \sigma_{am} = 0.01476, \sigma_{sm} = 0.17786\\ \sigma_{ms} &= 0.14918, \sigma_{as} = 0.01492, \sigma_{ss} = 0.19682\end{aligned}$$

The parameters α and Ψ_m are simply the capital cost share and scaling parameters associated with the production technology $\mathcal{F}^m\left(L_m, K_m\right)$. For the Cobb-Douglas technology we have the following estimated cost share expression

$$\hat{\alpha} = \frac{wl_m}{\left[p_m - \sum_{i=a,m,s} p_i\hat{\sigma}_{im}\right]Y_m} = \frac{wl_m}{C^m\left(w, r^k\right)Y_m} = \frac{wl_m}{wl_m + r^kK_m}$$

The estimated value of α is

$$\begin{aligned}\hat{\alpha}_m &= \frac{14129.7}{18273.5 + 14129.7} = 0.43606\\ \hat{\alpha}_s &= \frac{52733.4}{46895.7 + 52733.4} = 0.5293\end{aligned}$$

while the estimates of the manufacturing and service sector scaling parameters are:

$$\hat{\Psi}_m = \frac{TC_m}{\left(l_m\right)^{\hat{\alpha}_m}\left(\frac{r^kK_m}{r^k_{2001}}\right)^{1-\hat{\alpha}_m}} = \frac{97153.9}{\left(\frac{14129}{80268}\right)^{0.436}\left(\frac{18273}{0.1199}\right)^{1-0.436}} = 247.506$$

and the value-added technology for the manufacturing sector is

$$Y_m = 247.506\,(l_m)^{0.43606}\,(K_m)^{0.5639}$$

Straightforward substitutions yield the manufacturing cost function. It is left to the reader to derive the service sector cost function.

With Cobb-Douglas technologies, the land rental function is equal to

$$\pi^a(p_{va}, w, r^k)H = \phi_1^{\frac{\phi_1}{\phi_3}}\,\phi_2^{\frac{\phi_2}{\phi_3}}\,\phi_3\,(p_{va}\Psi_a)^{\frac{1}{\phi_3}}\,w^{\frac{-\phi_1}{\phi_3}}\,\left(r^k\right)^{\frac{-\phi_2}{\phi_3}}\,H$$

where $p_{va} = p_a(1-\sigma_{aa}) - p_m\sigma_{ma} + p_s\sigma_{sa}$. As above, estimates of the input-output coefficients are derived using the identity

$$\hat{\sigma}_{ia} \equiv \frac{p_i\sigma_{ia}Y_a}{TC_a}, \qquad i = a, m, s$$

where

$$\sigma_{ma} = 0.11262, \sigma_{aa} = 0.23299, \sigma_{sa} = 0.13567$$

Given the Cobb-Douglas technology

$$Y_a = \Psi_a\,(l_a)^{\phi_1}\,(K_a)^{\phi_2}\,H^{1-\phi_1-\phi_2}$$

the estimated values of ϕ_1 and ϕ_2 are given by:

$$\begin{aligned}
\hat{\phi}_1 &= \frac{wL_a}{\left[p_a - \sum_{j=a,m,s} p_j\hat{\sigma}_{ja}\right]Y_a} = \frac{wL_a}{wL_a + r^kK_a + \pi H} = 0.54045 \\
\hat{\phi}_2 &= \frac{r^kK_a}{\left[p_a - \sum_{j=a,m,s} p_j\hat{\sigma}_{ja}\right]Y_a} = \frac{r^kK_a}{wL_a + r^kK_a + \pi H} = 0.37986 \\
&1 - \hat{\phi}_1 - \hat{\phi}_2 = 0.07969
\end{aligned}$$

Given land is normalized to unity, $H = 1$, the estimate for the scaling parameter is

$$\hat{\Psi}_a = \frac{TC_a}{(l_a)^{\hat{\phi}_1}\left(\frac{r^kK_a}{r^k_{2001}}\right)^{\hat{\phi}_2} H^{1-\hat{\beta}_1-\hat{\beta}_2}} =$$

$$\frac{47818.9}{\left(\frac{13405.3}{80268.4}\right)^{0.54045}\left(\frac{9422.0}{0.11993}\right)^{0.37986}} = 1738.54$$

After tedious algebraic manipulations we get the agricultural value-added function:

$$\begin{aligned}\pi^a(p_{va}, w, r^k)H &= (1333.5p_a + 195.79p_m + 235.87p_s)^{12.549} \\ &\quad \times 0.00001216w^{-6.7823}\left(r^k\right)^{-4.767}\end{aligned}$$

The reader can verify that substituting $p_a = p_m = p_s = 1$ and $w = 80268.4$ and $r^k = 0.11993$ into the above expression yields

$$\pi^a\left(0.51872, 80268.4, 0.11993\right) = 1976.5$$

This is another example of using equilibrium conditions, or functions in the model, to check for scripting errors.

8.5 A three-sector SAM with composite capital and government

Chapter 6 presents a three-sector, open economy model with intermediate inputs, composite capital, and government. The government raises income with tariffs and indirect taxes, and uses this income to purchase goods and services and make transfer payments to households. As in Chapter 4, the economy employs three factors and intermediate inputs to produce three outputs.

Table 8.9 adds a government, denoted "Gov", sub-account, and a "Taxes" account to Table 8.7. Moving along the government row account, there is a single cell entry "Tax," the income received by the government from indirect taxes, tariffs, and households. Moving down the government column, the cell entries "DomCons$_i^g$" represents the value of government expenditures on final good $j = a, m, s$. At the bottom of this column is the entry "GovExp," the total value of the government's demand for goods and services.

Each entry down the "Taxes" column account represents a transfer of income from the taxes accounts to the recipient row account. Recall, entries along a row account represent receipts the row account receives from the intersecting column account: e.g., the entry "-ExportTax$_a$" is the payment the agricultural sector receives from the column account "Taxes" – a positive export tax, then is a subsidy. Then, moving down the "Taxes" column account, the entries "-ExportTax$_a$" and "-ExportTax$_m$" represent *receipts* from the "Taxes" account to the agricultural and manufacturing sectors: a positive value for either entry means, on net, the sector's exports were subsidized. The entry "Tax" is the payment from the "Taxes" account to the "Gov" account.

The "Taxes" row account has entries for indirect taxes and tariff income. Moving along the row account "Taxes", the cell entries "IndTax$_i$," $i = a, m, s$, represent the indirect taxes the government raises from sector i. The cell entries "Tariff$_a$" and "Tariff$_m$" represent the amount of income the government raises by taxing agricultural and manufacturing imports, respectively. Of course, negative values in any of these cells means on net, the sector benefitted from a government subsidy. The entry "Transfer" is a lump-sum value equal to the difference between government expenditures "GovExp" and the total value of indirect taxes and tariffs, adjusted for export taxes. Note that this value is computed as a residual in the SAM.

In the prior section, there was no distinction between domestic and world prices. Now, however, domestic prices are represented by p_j, $j = a, m, s$, while world prices are represented by p_j^w, $j = a, m$. More on this shortly.

8.5.1 Using the SAM to calibrate the empirical three-sector model with government and composite capital

Let τ_j^I denote the indirect tax rate paid by sector $j = a, m, s$, and let τ_j and τ_j^{EX} denote the tariff and export tax rates on

Table 8.9 Structure of a three-sector SAM with trade, intermediate inputs, composite capital and government

	Activity			Commodity			Factors			Agents	Taxes		Accum	Trade	Tot Receipts
	m	a	s	m	a	s	K	L	H	HH	Gov				
Activity:															
m				Q^c_m								- ExportTax$_m$		$p^w_m EX_m$	$p^w_m Y_m$
a					$p_a Q^{DA}_a$							- ExportTax$_a$		$p^w_a EX_a$	$p^w_a Y_a$
s						$p_s Y_s$									$p_s Y_s$
Commodity:															
m	$p_m Y_{m,m}$	$p_m Y_{m,a}$	$p_m Y_{m,s}$							DomCons^h_m	DomCons^g_m		Investment$_m$		$p_m Q^{DA}_m$
a	$p_a Y_{a,m}$	$p_a Y_{a,a}$	$p_a Y_{a,s}$							DomCons^h_a	DomCons^g_a		Investment$_a$		$p_a Q^{DA}_a$
s	$p_s Y_{s,m}$	$p_s Y_{s,a}$	$p_s Y_{s,s}$							DomCons^h_s	DomCons^g_s		Investment$_s$		$p_s Q_s$
Inputs															
K	$r^k K_m$	$r^k K_a$	$r^k K_s$												wL
L	wL_m	wL_a	wL_s												$r^k K$
H		πH													πT
Agents															
HH							$r^k K$	wL	πT						$G(\cdot)$
Gov												Tax			GovInc
Taxes	IndTax$_m$	IndTax$_a$	IndTax$_s$	Tariff$_m$	Tariff$_a$					Transfer					TaxSource
Accum										$G - p \cdot Q$					K
Trade				$p^w_m IM_m$	$p^w_a IM_a$										$\sum_{i=a,m,s} IM_i$
Tot Exp	TC_m	TC_a	TC_s	$p_m Y_m$	$p_a Y_a$	$p_s Y_s$	$r^k K$	wL	πT	$\mathcal{E}(\cdot) Q + K$	GovExp	TaxUse	K	$\sum_{i=a,m,s} EX_i$	

good $j = a, m$. Furthermore, define

$$\begin{aligned}
\text{IndTax}_j &\equiv \tau_j^I p_j Y_j, \quad j = a, m, s \\
\text{Tariff}_j &\equiv \tau_j p_j^w Y_j \quad j = a, m \\
-\text{ExportTax}_j &\equiv -\tau_j^{EX} p_j EX_j \quad j = a, m \\
\text{DomCons}_j^h &\equiv p_j Q_j^h \quad j = a, m, s \\
\text{DomCons}_j^g &\equiv p_i Q_j^g \quad j = a, m, s \\
\text{Investment}_j &= p_j Q_j^{inv} \quad j = a, m, s
\end{aligned}$$

As explained in Chapter 5, the government imposes a tax on households that is equal to the difference between tax revenue and government expenditures. Hence, total indirect taxes

$$REV_I \equiv \sum_{j=a,m,s} \tau_j^I p_j Y_j$$

plus total tariff income

$$REV_{Tar} \equiv \sum_{j=a,m} \tau_j p_j^w IM_j$$

less total government expenditures is equal to the value of this tax or transfer from household to the government:

$$\begin{aligned}
TR &\equiv REV_I + REV_{Tar} - \sum_{j=a,m,s} p_j Q_j^g \\
&= \sum_{j=a,m,s} \tau_j^I p_j Y_j + \sum_{j=a,m} \tau_j p_j^w IM_j - \sum_{j=a,m,s} p_j Q_j^g
\end{aligned}$$

This condition is a closure rule for the SAM with government and taxes, and is consistent with a balanced capital account and balanced government deficit.[6]

Table 8.10 presents the SAM underlying the empirical three-sector model of Chapter 6.

[6]The two-country models of Chapter 7 accomodate imbalanced capital accounts and government deficits.

Table 8.10 A three-sector SAM with trade, intermediate inputs, composite capital and government (trillion Turkish Lira)

	Activity			Commodity			Factors			Agents		Taxes	Accum	Trade	Tot Receipts
	m	a	s	m	a	s	K	L	H	HH	Gov				
Activity:															
m				110620.0											110620.0
a					42552.3									5304.1	47856.5
s						164341.2									164341.2
Commodity:															
m	46037.2	5385.3	23255.8							27084.0	1463.7		14322.0		117548.0
a	1433.5	11141.2	2326.6							26034.4	981.1	0.0	635.5		42552.3
s	17279.9	6488.6	30682.4							59540.7	35102.9		15246.6		164341.2
Inputs															
K	18273.5	9422.0	46895.7												74591.2
L	14129.7	13405.3	52733.4												80268.4
H		1976.5													1976.5
Agents															
HH	0.0	0.0	0.0				74591.2	80268.4	1976.5						156836.2
Gov	0.0	0.0	0.0									37547.5			37547.5
Taxes	13466.2	37.6	8447.3	1623.8						13972.6					37547.5
Accum										30204.2					30204.2
Trade				5304.1											5304.1
Tot Exp	110620.0	47856.5	164341.2	117548.0	42552.3	164341.2	74591.2	80268.4	1976.5	156835.9	37547.7	37547.5	30204.2	5304.1	

Consumption and production

Given the household consumption levels in Table 8.10, the felicity function is

$$u(Q_a, Q_m, Q_s) = Q_a^{0.2311} Q_m^{0.2516} Q_s^{0.5285}$$

and the corresponding expenditure function is

$$\mathcal{E}(p_a, p_m, p_s) Q \equiv 1.413 p_a^{0.2311} p_m^{0.2516} p_s^{0.5285} Q$$

To estimate the production cost shares and input-output coefficients embedded in Table 8.10 proceed as discussed above. To calculate the indirect tax rate, simply use the relationship

$$\hat{\tau}_j^I = \frac{\tau_j^I p_j Y_j}{TC_j}$$

where

$$\hat{\tau}_m^I = 0.12173, \hat{\tau}_a^I = 0.0007857 = 0.12173, \hat{\tau}_s^I = 0.0514$$

and the revised intermediate factor cost shares are

$$\begin{aligned}
\sigma_{mm} &= 0.41617, \sigma_{am} = 0.01296, \sigma_{sm} = 0.15621 \\
\sigma_{ma} &= 0.11253, \sigma_{aa} = 0.23280, \sigma_{sa} = 0.135585 \\
\sigma_{ms} &= 0.14151, \sigma_{as} = 0.01416, \sigma_{ss} = 0.1867
\end{aligned}$$

The reader can verify that with Cobb-Douglas technologies, given Table 8.10, the empirical manufacturing sector total cost function is

$$\begin{aligned}
TC_m &= (0.01296 p_a + 0.5379 p_m + 0.15621 p_s) Y_m + \\
&\quad 0.00801 \left(r^k\right)^{0.56394} w^{0.43606} Y_m
\end{aligned}$$

and the empirical service sector cost function is

$$\begin{aligned}
TC_s &= (0.01416 p_a + 0.1634 p_m + 0.2381 p_s) Y_s + \\
&\quad 0.004397 \left(r^k\right)^{0.4707} w^{0.5293} Y_s
\end{aligned}$$

while the empirical land rental function is

$$\pi^a(p_{va}, w, r^k)H = (1332.4p_a + 195.64p_m + 235.72p_s)^{12.549} \times 0.00001216 w^{-6.7823} \left(r^k\right)^{-4.767}$$

Composite capital is produced by combining final good output from manufacturing, agriculture, and services. Although other possible representations exist, we represent this process using a Cobb-Douglas technology. The minimum cost of producing a unit of composite capital is given by

$$c^k(p_m, p_a, p_s) \equiv$$

$$\min_{Y_{mk}, Y_{ak}, Y_{sk}} \left\{ \sum_{j=a,m,s}^{3} p_j Y_{jk} : 1 = (Y_{mk})^{\beta_{mk}} (Y_{ak})^{\beta_{ak}} (Y_{sk})^{\beta_{sk}} \right\}$$

where $\beta_{sk} = 1 - \beta_{mk} - \beta_{ak}$. The estimate of β_{jk} is given by

$$\hat{\beta}_{jk} = \frac{p_j Q_j^{inv}}{p_m Q_m^{inv} + p_a Q_a^{inv} + p_s Q_s^{inv}}$$

hence, from Table 8.10,

$$\hat{\beta}_{mk} = \frac{14322.0}{14322.0 + 635.5 + 15246.6} = 0.47417$$

$$\hat{\beta}_{ak} = \frac{635.5}{14322.0 + 635.5 + 15246.6} = 0.02104$$

$$\hat{\beta}_{mk} = \frac{15246.6}{14322.0 + 635.5 + 15246.6} = 0.50479$$

Then, given a Cobb-Douglas composite production technology and Table 8.10, the composite capital cost function is

$$c^k(p_m, p_a, p_s) = 2.18179 p_a^{0.02104} p_m^{0.47417} p_s^{0.50479}$$

Government expenditure and revenue

Chapter 6 introduces government. The government purchases goods from the agricultural, manufacturing, and service sectors, and pays for these goods with three sources of income: (i) lump-sum transfers from households, (ii) indirect taxes on sectoral production, and (iii) tariffs on traded goods. At each point in time, total government expenditure is assumed to be a constant share of GDP, $E^g = \lambda_g \left(r^k K + wL + \pi T\right)$. Similarly, government expenditure on sector $j = a, m, s$ is a constant share of total government expenditure, i.e., $p_j Q_j^g = \lambda_j^g E^g$. Here the estimate of λ_j^g is given by $\hat{\lambda}_j^g = \left(p_j Q_j^g\right) / E^g$, and the estimate of λ_g is given by $\hat{\lambda}_g = \left(p_m Q_m^g + p_a Q_a^g + p_s Q_s^g\right) / \left(r^k K + wL + \pi T\right)$. Direct calculations from Table 8.10 yield the following government expenditure shares:

$$\begin{aligned}
\hat{\lambda}_g &= \frac{1463.7 + 981.1 + 35102.9}{74591.2 + 80268.4 + 1976.5} = 0.23941 \\
\hat{\lambda}_m^g &= \frac{1463.7}{1463.7 + 981.1 + 35102.9} = 0.03898 \\
\hat{\lambda}_a^g &= \frac{981.1}{1463.7 + 981.1 + 35102.9} = 0.02613 \\
\hat{\lambda}_s^g &= \frac{35102.9}{1463.7 + 981.1 + 35102.9} = 0.93488
\end{aligned}$$

The three sources of government revenue each distort economic behavior in some way. Indirect taxes distort production via producers' zero profit conditions, but do not directly affect consumption behavior. Tariffs distort both production and consumption behavior: Production via the zero profit conditions and consumption via its impact on domestic consumption prices. On the other hand, lump-sum transfers introduce distortions via the flow budget constraint, but do not directly affect producer behavior.

Recall in Chapters 4 and 5, world and domestic prices were identical and normalized to unity, i.e.,

$$p_a = p_m = p_s = 1$$

With tariffs and export taxes, however, a wedge is driven between world and domestic prices. In such a case, world prices for agriculture and manufacturing satisfy

$$p_a = p_a^w \left(1 + \tau_a\right) = p_m = p_m^w \left(1 + \tau_m\right) = 1$$

Then, using tariff payment values, $\tau_i p_i^w IM_i$, and trade value data, $p_i^w IM_i$, an estimator of τ_j is:

$$\hat{\tau}_i = \frac{\tau_i p_i^w IM_i}{p_i^w IM_i}, \qquad i = m, s \tag{8.13}$$

From Table 8.10, then, our estimate of τ_m is

$$\hat{\tau}_m = \frac{1623.8}{5304.1} = 0.3061 \tag{8.14}$$

with corresponding world price equal to

$$\hat{p}_m^w = \frac{1}{1+\hat{\tau}_m} = \frac{1}{1.3061} = 0.7656$$

Note that $\hat{\tau}_m$, as calculated in (8.14), is not a unit tariff rate, as it is a value based on net trade values, not gross or inter-industry trade values. A more appropriate interpretation of $\hat{\tau}_i$ is that of a "tariff income multiplier," where at each point in time t, 1 Turkish lira in net imports generates $\hat{\tau}_i$ lira in tariff income.

8.6 Conclusion

The major objective of this chapter was to provide an introduction to social accounting matrices and share with the reader, part of our approach to transforming the theoretical models of Chapters 3, 4, 5, and 6 into their respective empirical specifications. This chapter focused on deriving numerical expressions for the primitives of each model, calculating the parameter values for these primitives, and then deriving the relevant indirect objective functions. The primitives of each model were the sectoral production functions and the household utility function, and in each model, the utility and production functions were specified as Cobb-Douglas functions. The other part of this transformation process is the numerical methods used to numerically solve the modeled economy. These issues are taken up in the next chapter.

In the case of the production function, the scale parameter and production cost shares were calculated directly from the input-output cost data in the activity column accounts of the social accounting matrix. The household consumption cost shares were calculated directly from the consumption data in the household column account of the social accounting matrix. Given the parameterized versions of each sector's technology, we then derived the cost function for the industry and service sectors, and the value-added function for agriculture. We also used the parameterized utility function of each model and derived the corresponding expenditure function.

One facet we address only slightly in this chapter is the use of a SAM to check for errors in programming and coding the numerical model. In other words, using the SAM to ensure the derivations of the empirical indirect objective functions and the parameter values substituted into those functions are correct. Imagine, for example, that the expenditure shares for a model have been calculated and the empirical expenditure function derived. Then, when Shepard's lemma is applied to the numerical expenditure function, and base year prices and GDP are substituted into the resulting numerical expression, the value of agricultural demand predicted by the numerical model should exactly equal the value of GDP spent on agriculture as given in the SAM. If this is not the case, then most likely there has been a mistake in estimating the consumption shares or there has been a mistake in deriving the expenditure function. The wise modeler will conduct such a test for each primitive and each indirect objective function in his or her model. In our experience, bypassing this step typically leads to "unexplained results" in subsequent simulations.

8.7 Appendix: Sector definitions

The sub-sectors to include in a sector depend upon the structure of the model. If the final good consumed is food, then the agricultural sector should include the sub-sectors that process and retail food. If agriculture is modeled as a primary sector, then its output is an intermediate input in those sectors of the economy that processed and distributed food. In either case, the structure of the model is dependent upon the choice of subsector aggregation into sectors.

The World Bank's World Development Indicators includes agriculture, industry with manufacturing identified as a subset of industry, and services as the three-sectors comprising a country's gross domestic product. Gross domestic product (GDP) at purchaser prices is the sum of gross value-added by all resident producers in the economy plus any product taxes (less subsidies) not included in the valuation of output. It is calculated without adjusting for depreciation of fabricated assets or for depletion and degradation of natural resources. As noted above, value-added is the net output of an industry after adding up all outputs and subtracting intermediate inputs. The industrial origin of value-added is determined by the International Standard Industrial Classification (ISIC) revision 3. Agriculture corresponds to ISIC divisions 1–5. This division includes crop and animal agriculture, hunting and related service activities, forestry, fishing and mining of coal. Food products is included in Manufacturing. Industry covers mining (other than coal), and manufacturing, construction, electricity, water, and gas (ISIC divisions 10–45). Manufacturing corresponds to industries belonging to ISIC divisions 15–37. The services sector correspond to ISIC divisions 50–99. This sector is derived as a residual (from GDP less agriculture and industry) and may not properly reflect the sum of service output, including banking and financial services. For some countries it includes product taxes (minus subsidies) and may also include statistical discrepancies.

Unfortunately, since food processing appears in the manufacturing sector, the ISIC 1–5 aggregation for agriculture is not an appropriate characterization of agricultural output if, in the modeled economy, some of the output is treated as a final good. Coal may more appropriately be included in the manufacturing sector. The GTAP database takes these aggregation issues into consideration. The list in Table A.1 provides a summary of the

Table 8.11 Sector aggregation in the GTAP data set

Agriculture	Manufacturing	Services
1. Grains	1. Forestry	1. Electricity
a. Rice	2. Fishing	2. Gas manufacture,
b. Wheat	3. Coal	distribution
c. Cereal grains	4. Oil	3. Water
2. Vegetables,	5. Gas	4. Construction
fruit and nuts	6. Minerals	5. Trade
3. Oil seeds	7. Textiles	6. Transport
4. Sugar cane,	8. Wearing apparel	7. Sea transport
sugar beets	9. Leather products	8. Air transport
5. Plant-based fibers	10. Wood products	9. Communication
6. Other crops	11. Paper products,	10. Financial services
7. Cattle, sheep,	publishing	11. Insurance
goats, horses	12. Petroleum,	12. Business services
8. Animal products	coal products	13. Recreation and
9. Raw milk	13. Chemical, rubber,	other services
10. Wool,	plastic products	14. Public
silk-worm cocoons	14. Mineral products	administration,
11. Meat: cattle, sheep	15. Ferrous metals	defence, health
goats, horses	16. Other metals	education
12. Meat products	17. Metal products	15. Dwellings
13. Vegetable oils	18. Motor vehicles	
and fats	and parts	
14. Dairy products	19. Transport equip.	
15. Processed rice	20. Electronic equip.	
16. Sugar	21. Machinery	
17. Food products	and equipment	
18. Beverages and	22. Other manufactures	
tobacco products		

Global Trade Analysis Data set organization. The GTAP sector aggregations include product taxes less subsidies, and depart from the values reported in the WDI. Of course, an economy's GDP as reported in the GTAP data base should correspond closely to the GDP values reported in the WDI data base.

9

Solution Methods in Transition Dynamics

This chapter introduces two numerical methods for solving recursive dynamic optimization problems in deterministic form: the time elimination method by Mulligan and Sala-i-Martin (1991, 1993) and the backward integration method by Brunner and Strulik (2002). Both procedures involve solving for transitional dynamics backwardly from a given initial value. Both methods improve upon the common procedure known as "shooting" (e.g. Judd, 1998, Chapter 10). Various other methods are also available, but they are not discussed here. Among these is a method developed by Trimborn et al. (2008) that, from our experience, has performed well on complex problems.[1] We present the key mathematical concepts underpinning the numerical method used to solve the models presented in the book, and in the Appendix to this chapter, outline the structure of the Mathematica code used to implement these concepts.

9.1 Time-elimination method

In solving dynamic optimization problems such as the Ramsey growth model with one state variable, we make use of a numerical solution method called the time elimination method introduced in Mulligan and Sala-i-Martin (1991, 1993). The time elimination method essentially involves transforming a boundary problem into an initial value problem as explained below, rendering the transition path solution to our dynamic optimization problem rather manageable.

Consider the system of autonomous differential equations that

[1] An earlier version of this method performed well for solving a continuous time – two state variable problem in a study by Gaitán and Roe (2005).

T.L. Roe et al., *Multisector Growth Models*, DOI 10.1007/978-0-387-77358-2_9,

help characterize equilibria in a two-sector model of an economy as given in Equations (3.60) and (3.62). Express these equations as

$$\dot{k}(t) = g^k(k(t), p(t)) \tag{9.1}$$

$$\dot{p}(t) = g^p(k(t), p(t)) \tag{9.2}$$

where $k(t)$ is the state variable through time, and $p(t)$ is the control variable. As explained in Mulligan and Sala-i-Martin (1991), this system is a *boundary value*-type problem because $k(t)$ and $p(t)$ are solutions to this system of equations under the boundary conditions $k(0)$ and transversality condition (TVC) that specify the value of $k(t)$ at each point in time. Given the initial condition $k(0)$, one can guess the initial control $p(0)$ and examine the subsequent dynamics of the economy given by the system of differential equations to see if the TVC's are violated. If so, one must revise the initial guess and repeat. This particular procedure to solve the system is called "shooting", which may prove to be conceptually as well as computationally inefficient and difficult. However, a boundary value problem can be represented as an initial value problem, that is, if the initial value of the control variable p corresponding to the initial value of k were known, then simply integrating (9.1) and (9.2) would yield the optimal solutions $k(t)$ and $p(t)$ at each point in time.

By transforming a boundary-value problem into an initial-value problem, the time elimination method expresses the control variable as a function of the state variable, instead of as a function of time. That is, a *policy function*

$$p = P(k) \tag{9.3}$$

that solves for the optimal values for the control variable p as a function of the state variable k is assumed to exist. If this function exists, then (9.3) can be substituted into (9.1) to obtain

$$\dot{k}(t) = g^k(k(t), P(k(t))) \tag{9.4}$$

Then, given the initial condition $k(0)$, one could simply integrate (9.4) forward to a known value, the steady-state k^{ss}, and thus obtain the optimal values $\{k(t)\}_{t\in[o,\infty)}$. Once the optimal values $k(t)$ are known, $\{p(t)\}_{t\in[0,\infty)}$ series can be formed based on the *policy function.* Hence, the challenge here is to define or, describe the policy function.

One possible way of obtaining a policy function is to draw the *phase diagram* of the system and identify the stable arm that contains stable equilibria along the transition path. For example, consider the standard one-sector Ramsey growth model in which the equilibria are characterized by the following differential equations:[2]

$$\begin{aligned} \dot{k} &= f(k) \quad (x \mid n \mid \delta)k \quad c \\ \frac{\dot{c}}{c} &= \frac{1}{\theta}\left[f'(k) - \delta - \rho - \theta x\right] \end{aligned}$$

where k is a state variable, c is a control variable.

The phase diagram in Figure 9.1 depicts the saddle-path stability of the system, in which there is a stable arm (trajectory that converges towards the stable equilibrium point), and an

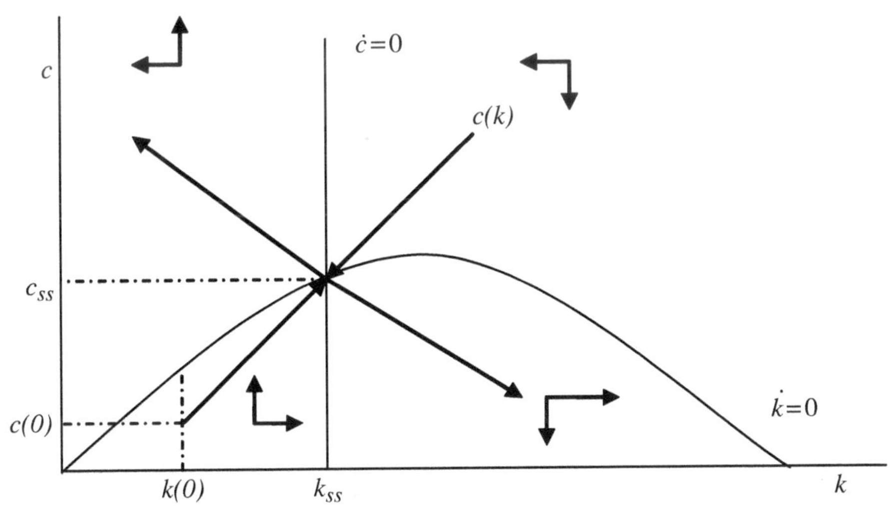

Figure 9.1 Phase diagram of the one-sector Ramsey growth model

[2] For the full treatment of the one-sector Ramsey growth model, see Barro and Sala-i-Martin (2004).

unstable arm (trajectory that diverges away from the stable equilibrium point) and specifies the four possible quadrants in which the economy may evolve. If the economy starts life at the upper left quadrant, then the dynamics are such that the k and c values move in northwest direction and move farther away from the steady-state equilibrium. If the economy starts life at the lower right quadrant, the c falls but k increases, so again the path moves farther away from the steady-state equilibrium. If the initial condition in the economy is $k(0)$ and the corresponding $c(0)$ is guessed correctly in the lower left quadrant, then the economy follows the path towards the steady-state equilibrium. If $c(0)$ is chosen too low or too high, again the economy diverges away from the equilibrium. Here, the stable arm or the function $c(k)$ which shows the trajectory of equilibrium solutions for the pair (k, c) is also known as the policy function.

In many cases, particularly in systems with more than one sector, the algebraic expression or closed form representation of the stable arm may not be available. Secondly, even if one works out the graphical representation of the model correctly, it does not provide a quantitative evaluation of the model. For the more general case, the stable arm (or plane), if it exists, can be found by linearizing the system of differential equations in the neighborhood of the known steady-state, and then by verifying that the determinant of the characteristic matrix is negative. That is, one can verify the saddle path stability of the system by linearizing the system of differential equations around the known steady-state. For a system of two differential equations, this requires that one of the two eigenvalues is negative, and the other positive, implying that the system has saddle-path stability. The eigenvector associated with the negative eigenvalue then corresponds to the stable arm (or the saddle-path) as further discussed in Section 9.1.2 below. If the system of differential equations is an $n \times n$ system with a single state variable, then for saddle-path stability, one requires one negative eigenvalue and (n-1) positive eigenvalues.[3]

[3] On the other hand, if it is a system of m state variables, then m negative eigenvalues

Returning to our model in (9.1) and (9.2), we recognize that the ratio of the time derivative of the control variable, $\dot{p}$, to the time derivative of the state variable, $\dot{k}$, is the first order derivative of the policy function, i.e.,

$$\begin{aligned} \dot{p}(t) &= \frac{dP(k(t))}{dk(t)}\dot{k}(t) \\ \frac{\dot{p}(t)}{\dot{k}(t)} &= \frac{dP(k(t))}{dk(t)} \end{aligned}$$

Essentially, we know from (9.1) and (9.2) that

$$\frac{\dot{p}}{\dot{k}} = \frac{g^p(k, P(k))}{g^k(k, P(k))}$$

Hence, it must be the case that

$$\frac{dP(k)}{dk} = \frac{g^p(k, P(k))}{g^k(k, P(k))} \tag{9.5}$$

Equation (9.5) is the slope of the policy function with no argument in time t. We have eliminated *time* from the system. Therefore, the TVC's are no longer asymptotic boundary conditions. Since it is typically not possible to obtain closed form solutions, resorting to a numerical solution is necessary. Even though we do not know the pair $[k(0), p(0)]$, we know that the policy function goes through the steady-state point (k^{ss}, p^{ss}). Starting from this known point and then solving (9.5) i.e., integrating backwards, enables us to determine the rest of the policy function.

However one important problem arises; the slope of the policy function at the steady-state is indeterminate, i.e.

$$\frac{dP(k)}{dk}\Big|_{k\cong k^{ss}} = \frac{g^p(k^{ss}, P(k^{ss}))}{g^k(k^{ss}, P(k^{ss}))} = \frac{0}{0} \tag{9.6}$$

The problem of indeterminacy prevents us from tracing out the rest of the policy function. However, we can circumvent this

---

are needed to satisfy saddle-path stability condition, and the policy function is no longer a curve, but rather an m-dimensional plane.

problem by determining the slope of the policy function at the steady-state. There are two procedures to follow in order to obtain the slope of the policy function at the steady-state. One is the L'Hopital's rule approach, the other is the Eigenvalues-Eigenvectors (EE) approach.[4] Below, each procedure is presented in detail.

9.1.1 L'Hopital's rule approach

Applying L'Hopital's rule[5] to (9.6) allows us to compute the slope of the policy function at the steady-state which is otherwise undefined. Employing L'Hopital's rule in (9.6), we have

$$\frac{dP(k)}{dk}\Big|_{k\cong k^{ss}}=\frac{\frac{\partial g^p}{\partial k}+\frac{\partial g^p}{\partial p}\frac{dP(k)}{dk}}{\frac{\partial g^k}{\partial k}+\frac{\partial g^k}{\partial p}\frac{dP(k)}{dk}}\Big|_{k\cong k^{ss}}$$

then,

$$\frac{dP(k)}{dk}\left[\frac{\partial g^k}{\partial k}+\frac{\partial g^k}{\partial p}\frac{dP(k)}{dk}\right]\Big|_{k\cong k^{ss}}=\left\{\frac{\partial g^p}{\partial k}+\frac{\partial g^p}{\partial p}\frac{dP(k)}{dk}\right\}\Big|_{k\cong k^{ss}}$$

or,

$$\left(\frac{dP(k)}{dk}\right)^2\frac{\partial g^k}{\partial k}+\left(\frac{\partial g^k}{\partial k}-\frac{\partial g^p}{\partial k}\right)\frac{dP(k)}{dk}-\frac{\partial g^p}{\partial k}=0 \qquad (9.7)$$

Here, the slope $dP(k)/dk$ is treated as an unknown variable, whereas $\partial g^k/\partial k$ and $\partial g^p/\partial k$ are evaluated at the steady-state, and hence are real numbers. Essentially, the quadratic form (9.7) has two roots, i.e., two solutions to $dP(k)/dk$. These two solutions correspond to the slopes of the two trajectories passing

[4] The EE approach is used in all of the empirical examples presented in the book because of the flexibility of the Mathematica software.

[5] Suppose f and g are differentiable in an interval (α,β) around a, except possibly at a, and suppose that $f(x)$ and $g(x)$ both approach 0 when $x\to a$. If $g'(x)\neq 0$ for all $x\neq a$ in (α,β) and $\lim_{x\to a} f'(x)/g'(x)=L$ (L finite, $L=\infty$ or $L=-\infty$), then

$$\lim_{x\to a}\frac{f(x)}{g(x)}=\lim_{x\to a}\frac{f'(x)}{g'(x)}=L$$

(Sydsæter et al., 1999).

through the steady-state. In fact, one of the trajectories is associated with the stable arm, the other the unstable arm. But the problem here is we may not have enough information as to which solution corresponds to the stable arm of the system so that a trial and error numerical procedure is often required. This procedure can be time consuming for complex systems of equations.

9.1.2 Eigenvalues-eigenvectors approach

Suppose that we have a system of differential equations given by

$$\dot{y}(t) = Ay(t) \tag{9.8}$$

where A is an $n \times n$ square matrix which is not necessarily diagonal. For instance, the matrix A can be thought of as the Jacobian obtained from linearizing (9.1) and (9.2) around the steady-state values of k^{ss} and p^{ss}. One procedure that can be employed to solve this system of differential equations is the diagonalization of the matrix A. The diagonal elements in the diagonalized matrix are the eigenvalues, and the signs of these eigenvalues determine the stability of the system of differential equations. Particularly in a 2×2 system, if there are two eigenvalues that are real with opposite signs, the system is said to be saddle-path stable, and the stable arm corresponds to the eigenvector associated with the negative eigenvalue on the diagonal. The unstable arm is represented by the eigenvector associated with the positive eigenvalue.

Given the $n-$dimensional square matrix A, we can find the values of a scalar λ and the corresponding non-zero column vectors v such that

$$(A - \lambda I)v = 0 \tag{9.9}$$

where I is the $n \times n$ identity matrix. The system of homogeneous linear equations has non-trivial solutions if and only if

$$\det(A - \lambda I) = 0$$

The equation $\det(A - \lambda I) = 0$ is called the characteristic equation of A, and its roots $\lambda_1, \lambda_2, ..., \lambda_n$ are the characteristic roots or the eigenvalues of A. Note that by (9.9), each eigenvalue λ_i is associated with a vector v_i that satisfies

$$Av_i = \lambda_i v_i$$

in which the $n \times 1$ vector v_i is the characteristic vector, or the eigenvector. These n, $n \times 1$ column vectors can be rearranged to write

$$AV = VD$$

where D is an $n \times n$ diagonal matrix with diagonal elements of eigenvalues λ_i and V is a $n \times n$ matrix of eigenvectors.

If $\det(V) \neq 0$ so that the eigenvectors are linearly independent, we can write

$$V^{-1}AV = D$$

That is, if the matrix A is premultiplied with the inverse of V and postmultiplied by V, we get a diagonal matrix transformation D with eigenvalues as its diagonal elements.

Now we define the variables $z(t)$ as follows:

$$z(t) = V^{-1}y(t)$$

Since V^{-1} is a matrix of constants, one can rewrite

$$\begin{aligned} \dot{z}(t) &= V^{-1}\dot{y}(t) \\ &= V^{-1}Ay(t) \\ &= V^{-1}AVV^{-1}y(t) \\ &= Dz(t) \end{aligned}$$

This is a system of n one-dimensional differential equations:

$$\begin{aligned} \dot{z}_1(t) &= \lambda_1 z_1(t) \\ \dot{z}_2(t) &= \lambda_2 z_2(t) \\ &\vdots \\ \dot{z}_n(t) &= \lambda_n z_n(t) \end{aligned}$$

Considering a simple 2×2 system, the solution can be expressed as

$$\begin{bmatrix} z_1 \\ z_2 \end{bmatrix} = \begin{bmatrix} e^{\lambda_1 t} & 0 \\ 0 & e^{\lambda_2 t} \end{bmatrix} \begin{bmatrix} b_1 \\ b_2 \end{bmatrix}$$

where the b_i are arbitrary constants of integration. Using $y = Vz$, one can transform the solution for z variables back to the original y variables:

$$\begin{bmatrix} y_1 \\ y_2 \end{bmatrix} = \begin{bmatrix} v_{11} & v_{12} \\ v_{21} & v_{22} \end{bmatrix} \begin{bmatrix} z_1 \\ z_2 \end{bmatrix}$$

or,

$$\begin{aligned} y_1(t) &= v_{11}z_1 + v_{12}z_2 \\ &= v_{11}e^{\lambda_1 t}b_1 + v_{12}e^{\lambda_2 t}b_2 \\ y_2(t) &= v_{21}z_1 + v_{22}z_2 \\ &= v_{21}e^{\lambda_1 t}b_1 + v_{22}e^{\lambda_2 t}b_2 \end{aligned}$$

Recall that the goal was to solve the system of the two differential equations in the neighborhood of the steady-state. In order for the system to be saddle-path stable, we require the signs of the eigenvalues λ_1 and λ_2 to be opposite. Let λ_1 be the negative eigenvalue. In that case, the stable arm will correspond to the eigenvector $\begin{bmatrix} v_{11} & v_{21} \end{bmatrix}'$ associated with λ_1, allowing us to ignore the unstable arm corresponding to the eigenvector associated with λ_2. Then, setting $b_2 = 0$ to ignore the influence of the unstable root, we can solve for the functional form of the stable arm as

$$y_2 = f(y_1)$$

assuming that on the phase diagram, y_1 is located on the x-axis, y_2 is located on the y-axis. Then, solving for b_1 in y_1 and plugging it back in y_2, the stable arm can be found as

$$y_2 = f(y_1) = \left(\frac{v_{21}}{v_{11}}\right) y_1$$

Now going back to our original dynamic optimization problem in Chapter 3, recall that we have the following 2×2 system of differential equations that characterize the equilibria, which above we refer to as (9.1) and (9.2). To be able to present this system in the form of (9.8), i.e. to linearize the system, a first-order Taylor series expansion around the steady state is employed:[6]

$$g^k(k,p) \cong$$

$$g^k(k^{ss},p^{ss}) + (\partial g^k/\partial k) \mid_{(k^{ss},p^{ss})} \times (k - k^{ss}) + (\partial g^k/\partial p) \mid_{(k^{ss},p^{ss})} \times (p - p^{ss})$$

$$g^p(k,p) \cong$$

$$g^p(k^{ss},p^{ss}) + (\partial g^p/\partial k) \mid_{(k^{ss},p^{ss})} \times (k - k^{ss}) + (\partial f_2/\partial p) \mid_{(k^{ss},p^{ss})} \times (p - p^{ss})$$

Since

$$g^k(k^{ss},p^{ss}) = g^p(k^{ss},p^{ss}) = 0$$

we simply have

$$g^k(k,p) \cong$$

$$(\partial g^k/\partial k) \mid_{(k^{ss},p^{ss})} \times (k - k^{ss}) + (\partial g^k/\partial p) \mid_{(k^{ss},p^{ss})} \times (p - p^{ss})$$

$$g^p(k,p) \cong$$

$$(\partial g^p/\partial k) \mid_{(k^{ss},p^{ss})} \times (k - k^{ss}) + (\partial g^p/\partial p) \mid_{(k^{ss},p^{ss})} \times (p - p^{ss})$$

In 2×2 matrix form, we can write the system of differential equations above as,

$$\begin{bmatrix} \dot{\tilde{k}} \\ \dot{\tilde{p}} \end{bmatrix} = \begin{bmatrix} (\partial g^k/\partial k) & (\partial g^k/\partial p) \\ (\partial g^p/\partial k) & (\partial g^p/\partial p) \end{bmatrix} \mid_{(k^{ss},p^{ss})} \times \begin{bmatrix} \tilde{k} \\ \tilde{p} \end{bmatrix}$$

where we have performed a change of variables so that

$$\begin{aligned} \tilde{k} &= k - k^{ss} \\ \tilde{p} &= p - p^{ss} \end{aligned}$$

Essentially, to replicate the notation in (9.8), we now redefine

$$\dot{y} \equiv \begin{bmatrix} \dot{\tilde{k}} \\ \dot{\tilde{p}} \end{bmatrix}$$

---

[6] These and all calculations below are performed by coding in Mathematica.

and

$$y \equiv \begin{bmatrix} \tilde{k} \\ \tilde{p} \end{bmatrix}$$

with the Jacobian matrix A,

$$A \equiv \begin{bmatrix} (\partial g^k/\partial k) & (\partial g^k/\partial p) \\ (\partial g^p/\partial k) & (\partial g^p/\partial p) \end{bmatrix} |_{(k^{ss},p^{ss})}$$

so that the general solution is given by

$$\begin{aligned} \tilde{k}(t) &= v_{11}e^{\lambda_1 t}b_1 + v_{12}e^{\lambda_2 t}b_2 \\ \tilde{p}(t) &= v_{21}e^{\lambda_1 t}b_1 + v_{22}e^{\lambda_2 t}b_2 \end{aligned}$$

Picking λ_1 as the negative eigenvalue and ignoring the unstable root by setting $b_2 = 0$,

$$\begin{aligned} \tilde{k}(t) &= v_{11}e^{\lambda_1 t}b_1 \\ \tilde{p}(t) &= v_{21}e^{\lambda_1 t}b_1 \end{aligned}$$

Taking time derivatives, we have[7]

$$\dot{k}(t) = \lambda_1 v_{11}e^{\lambda_1 t}b_1 \tag{9.10}$$

$$\dot{p}(t) = \lambda_1 v_{21}e^{\lambda_1 t}b_1 \tag{9.11}$$

Under the time elimination method, we have already argued in (9.5) that

$$\frac{\dot{p}}{\dot{k}} = \frac{g^p(k,P(k))}{g^k(k,P(k))} = \frac{dP(k)}{dk} \tag{9.12}$$

is the slope of the policy function, however indeterminate at the steady-state. Then, using (9.10) and (9.11), the slope of the policy function evaluated at (or in the neighborhood of) the steady-state can be found as

$$\frac{\dot{p}}{\dot{k}} = \frac{g^p(k,P(k))}{g^k(k,P(k))} |_{k \cong k^{ss}} = \frac{\lambda_1 v_{21}e^{\lambda_1 t}b_1}{\lambda_1 v_{11}e^{\lambda_1 t}b_1} = \frac{v_{21}}{v_{11}}$$

---

[7] Note that since $\tilde{k} = k - k^{ss}$, $\tilde{p} = p - p^{ss}$, $\dot{\tilde{k}} = \dot{k}$ and $\dot{\tilde{p}} = \dot{p}$.

Knowing the slope at the steady-state (which is the initial value), now integrate (9.12) backward towards $k(0)$ with respect to k to obtain the policy function. One can do this since the functions $g^k(.)$ and $g^p(.)$ are known. Having obtained these $P(k)$ values under the policy function, we solve the differential equation

$$\dot{k}(t) = g^k(k(t), P(k(t)))$$

as a function of time and obtain the sequence

$$k(t) = \mathtt{K(t)}$$

With the solution to the sequence $k(t)$, we return to the policy function $P(k)$ to construct the sequence for p

$$p(t) = P(\mathtt{K}(t))$$

This solution procedure is easily generalized to the one-state-variable and multiple control variables case. Consider the two country world model presented in Chapter 7 for the case of capital mobility between countries. The inter-temporal equations for this model are reduced to three differential equations obtained by solving (7.32), (7.35) and (7.36) for $\dot{\bar{k}}^w$, $\dot{p}_a$ and $\dot{p}_s$ as a function of $\bar{k}^w, p_a$, and p_s. Denote this solution by

$$\begin{aligned}
\dot{\bar{k}}^w &= f^k\left(\bar{k}^w, p_a, p_s\right) \\
\dot{p}_a &= f^a\left(\bar{k}^w, p_a, p_s\right) \\
\dot{p}_s &= f^s\left(\bar{k}^w, p_a, p_s\right)
\end{aligned}$$

Then, we will be looking for two policy functions rather than one,

$$\begin{aligned}
\dot{p}_a &= P^a(\bar{k}^w) \\
\dot{p}_s &= P^s(\bar{k}^w)
\end{aligned}$$

with the following slopes

$$\frac{\dot{p}_a}{\dot{\bar{k}}^w} = \frac{dP^a(\bar{k}^w)}{d\bar{k}^w} = \frac{f^a(\bar{k}^w, p_a, p_s)}{f^k(\bar{k}^w, p_a, p_s)} \tag{9.13}$$

$$\frac{\dot{p}_s}{\dot{\bar{k}}^w} = \frac{dP^s(\bar{k}^w)}{d\bar{k}^w} = \frac{f^s(\bar{k}^w, p_a, p_s)}{f^k(\bar{k}^w, p_a, p_s)} \tag{9.14}$$

Redefining $\dot{y}$ and y to accommodate for two control variables $\dot{p}_a$ and $\dot{p}_s$ and one state variable $\bar{k}^w$, we have

$$\dot{y} \equiv \begin{bmatrix} \dot{\tilde{k}}^w \\ \dot{\tilde{p}}_a \\ \dot{\tilde{p}}_s \end{bmatrix}$$

$$y \equiv \begin{bmatrix} \tilde{k}^w \\ \tilde{p}_a \\ \tilde{p}_s \end{bmatrix}$$

with the Jacobian B,

$$B \equiv \begin{bmatrix} (\partial f^k/\partial \bar{k}^w) & (\partial f^k/\partial p_a) & (\partial f^k/\partial p_s) \\ (\partial f^a/\partial \bar{k}^w) & (\partial f^a/\partial p_a) & (\partial f^a/\partial p_s) \\ (\partial f^s/\partial \bar{k}^w) & (\partial f^s/\partial p_a) & (\partial f^s/\partial p_s) \end{bmatrix} |_{(\bar{k}^{w,ss}, p_a^{ss}, p_s^{ss})}$$

which yields

$$\begin{aligned} \tilde{k}^w(t) &= v_{11}e^{\lambda_1 t}b_1 + v_{12}e^{\lambda_2 t}b_2 + v_{13}e^{\lambda_3 t}b_3 \\ \tilde{p}_a(t) &= v_{21}e^{\lambda_1 t}b_1 + v_{22}e^{\lambda_2 t}b_2 + v_{23}e^{\lambda_3 t}b_3 \\ \tilde{p}_s(t) &= v_{31}e^{\lambda_1 t}b_1 + v_{32}e^{\lambda_2 t}b_2 + v_{33}e^{\lambda_3 t}b_3 \end{aligned}$$

where λ_1, λ_2, and λ_3 are the eigenvalues. For saddle-path stability, this system is required to have one eigenvalue with negative real parts and two eigenvalues with positive real parts. Assuming that λ_1 is the negative eigenvalue, ignoring the influence of the unstable roots ($b_2 = b_3 = 0$), and following the same procedure as we followed in the one-control variable case, we find that the slopes of the policy functions at the steady-state are given as

$$\begin{aligned} \frac{\dot{p}_a}{\dot{k}} &= \frac{f^a(\bar{k}^w, P(\bar{k}^w), C(\bar{k}^w))}{f^k(\bar{k}^w, P(\bar{k}^w), C(\bar{k}^w))} |_{\bar{k}^w \simeq \bar{k}^{w,ss}} = \frac{\lambda_1 v_{21} e^{\lambda_1 t} b_1}{\lambda_1 v_{11} e^{\lambda_1 t} b_1} = \frac{v_{21}}{v_{11}} \\ \frac{\dot{p}_s}{\dot{\bar{k}}^w} &= \frac{f^s(\bar{k}^w, P(\bar{k}^w), C(\bar{k}^w))}{f^k(\bar{k}^w, P(\bar{k}^w), C(\bar{k}^w))} |_{\bar{k}^w \simeq \bar{k}^{w,ss}} = \frac{\lambda_1 v_{31} e^{\lambda_1 t} b_1}{\lambda_1 v_{11} e^{\lambda_1 t} b_1} = \frac{v_{31}}{v_{11}} \end{aligned}$$

Obviously,

$$\frac{dP^a(\bar{k}^w)}{d\bar{k}^w} = \frac{v_{21}}{v_{11}} \quad (9.15)$$

$$\frac{dP^s(\bar{k}^w)}{d\bar{k}^w} = \frac{v_{31}}{v_{11}} \quad (9.16)$$

can be used to obtain the values for p_a and p_s corresponding to the interval $[k(0), k_{ss}]$. Having solved for these values, the differential equation

$$\dot{\bar{k}}^w(t) = f^k(\bar{k}^w(t), P^a(\bar{k}^w(t)), P^s(\bar{k}^w(t)))$$

can now be solved to obtain the sequence $\{k(t)\}_{t\in[0,\infty)}$, which can then be simply plugged in each of the policy functions to generate the sequences $\{p_a(t)\}_{t\in[0,\infty)}$ and $\{p_s(t)\}_{t\in[0,\infty)}$.

9.1.3 Mathematica code

In spite of the straightforward procedures described in the preceding section, the implementation may not appear as straightforward. Here we follow Mathematica syntax to show how easily these procedures can be implemented.

The single policy function case

Suppose that we have a 2×2 differential equations system summarized by the Jacobian A, given as

$$\begin{bmatrix} \dot{k} \\ \dot{p} \end{bmatrix} = A \begin{bmatrix} k \\ p \end{bmatrix}$$

with

$$A = \begin{bmatrix} a_{11} & a_{12} \\ a_{21} & a_{22} \end{bmatrix}$$

and

$$\dot{k} = g^k(k, p)$$
$$\dot{p} = g^p(k, p)$$

Using Mathematica, the numerical values of the elements of A can be calculated as follows:

$$\begin{aligned}
a_{11} &= \mathtt{D}\left[g^k(k,p),k\right]/.\{k \to k^{ss}, p \to p^{ss}\} \\
a_{12} &= \mathtt{D}\left[g^k(k,p),p\right]/.\{k \to k^{ss}, p \to p^{ss}\} \\
a_{21} &= \mathtt{D}\left[g^p(k,p),k\right]/.\{k \to k^{ss}, p \to p^{ss}\} \\
a_{22} &= \mathtt{D}\left[g^p(k,p),p\right]/.\{k \to k^{ss}, p \to p^{ss}\}
\end{aligned}$$

Here, for example the a_{11} element of the Jacobian denotes the first order derivative "$\mathtt{D}[\cdot]$" of the function $g^k(\cdot)$ with respect to its first argument, k, evaluated at the steady-state pair (k^{ss}, p^{ss}), denoted by $\{k \to k^{ss}, p \to p^{ss}\}$. Here, $\mathtt{D}[\cdot]$ is the first order derivative operator in Mathematica. Suppose now that numerical solution to our problem yields

$$A = \begin{bmatrix} -0.869172 & -78754.3 \\ 9.85648 & 0.899172 \end{bmatrix}$$

In Mathematica, the eigenvalues operator Eigenvalues$[A]$ is used to obtain the eigenvalues of the matrix A. These values are found as 0.892936 and -0.059294. In addition, the eigenvectors vector operator in Mathematica is Eigenvectors$[A]$. The eigenvectors of the matrix A are

$$\begin{aligned}
\begin{bmatrix} v_{11} \\ v_{21} \end{bmatrix} &= \begin{bmatrix} 1.0 \\ -0.00001217 \end{bmatrix} \\
\begin{bmatrix} v_{12} \\ v_{22} \end{bmatrix} &= \begin{bmatrix} -1.0 \\ 0.000010284 \end{bmatrix}
\end{aligned}$$

We now choose the negative eigenvalue, and pick the eigenvector associated with the negative eigenvalue for saddle-path stability of the system. Ignoring the unstable root, we have the solution to the differential equations system given by

$$\begin{aligned}
k(t) &= (-1.0)e^{-0.059294t}b_2 \\
p(t) &= (0.000010284)e^{-0.059294t}b_2
\end{aligned}$$

which implies

$$\begin{aligned}
\dot{k}(t) &= (0.059294)e^{-0.059294t}b_2 \\
\dot{p}(t) &= (-0.059294)(0.000010284)e^{-0.059294t}b_2
\end{aligned}$$

Recall that the slope of the policy function is given by

$$\frac{\dot{p}}{\dot{k}} = \frac{dP(k)}{dk}$$

we have this slope evaluated at the steady-state as

$$\begin{aligned}\frac{\dot{p}}{\dot{k}} &= \frac{(-0.059294)(0.000010284)e^{-0.059294t}b_2}{(0.059294)e^{-0.059294t}b_2} \\ &= -0.000010284\end{aligned}$$

which is essentially the ratio of the two elements in the eigenvector associated with the negative eigenvalue, i.e. v_{22}/v_{12}.

Now that we have obtained the slope of the policy function evaluated at the steady state, we can integrate the policy function *backwards* from the steady state p^{ss} (known) and obtain the p values in the interval $[k(0), k^{ss}]$ where both of $k(0)$ and k^{ss} are known values. This can be accomplished in Mathematica with the `NDSolve` operator. The syntax is:

$$\begin{aligned} s &= \texttt{NDSolve}[\{P'[k]{==}\texttt{If}[k{==}k^{ss}, \texttt{slope}, (g^p[k, P[k]]/g^k \\ &\quad [k, P[k]])], P[k^{ss}]{==}p^{ss}\}, P, \{k, k(0), k^{ss}\}] \end{aligned} \tag{9.17}$$

Note that we have embedded the operator "$\texttt{If}[k == k^{ss}, \texttt{slope}$" inside in the `NDSolve` operator.

The solution is the sequence p over the range $[k(0), k^{ss}]$. Knowing this sequence, now one can recall the differential equation

$$\dot{k}(t) = g^k(k(t), P(k(t)))$$

and solve it for $k(t)$, for all $t \in [0, T], T < \infty$, *given* the solution (9.17) as follows[8]:

$$\begin{aligned} s2 &= \texttt{NDSolve}[\{k'[t]{==}g^k(k[t], P[k[t]]/.s), k[t], \\ &\quad k[0]{==}k(0)\}, k, \{t, 0, T\}] \end{aligned}$$

[8] To solve backward in time from the point $t = 0$ of fitting the model to data, simply set $t \in [-\text{value}, T]$ where "value" is the backward point in time one desires the model to solve to.

The syntax "/.s" simply recalls the values of p from (9.17). Hence, we now have a time series for the optimal solution $\{k(t)\}_{t\in[0,T]}$ for large T ($T \to \infty$ at the limit),which is "stored" in the register $s2$. Once the time series $\{k(t)\}_{t\in[0,\infty)}$ is obtained, now it is straightforward to construct the optimal sequence $\{p(t)\}_{t\in[0,}$ from

$$p(t) = P(k(t))_{t\in[0,\infty)}$$

using the syntax

$$p[t\_] = (p[k[t]]/.s2/.s)$$

The value of the model's remaining endogenous variables are obtained by substituting $\{k\left(t\right), p(t)\}_{t\in[0,\infty)}$ into the reduced forms obtained from the intra-temporal equilibrium conditions.

The two policy function case

In Mathematica syntax, the Jacobian for the two policy function - one state variable case of the mobile capital model of Chapter 7 is the following 3×3 matrix:

$$\begin{aligned}
a_{11} &= \mathtt{D}\left[f^k\left(\bar{k}^w, p_a, p_s\right), \bar{k}^w\right]/.\left\{\bar{k}^w \to \bar{k}^{w,ss}, p_a \to p_a^{ss}, p_s \to p_s^{ss}\right\} \\
a_{12} &= \mathtt{D}[f^k\left(\bar{k}^w, p_a, p_s\right), p_a]/.\left\{\bar{k}^w \to \bar{k}^{w,ss}, p_a \to p_a^{ss}, p_s \to p_s^{ss}\right\} \\
a_{13} &= \mathtt{D}\left[f^k\left(\bar{k}^w, p_a, p_s\right), p_s\right]/.\left\{\bar{k}^w \to \bar{k}^{w,ss}, p_a \to p_a^{ss}, p_s \to p_s^{ss}\right\} \\
a_{21} &= \mathtt{D}\left[f^a\left(\bar{k}^w, p_a, p_s\right), \bar{k}^w\right]/.\left\{\bar{k}^w \to \bar{k}^{w,ss}, p_a \to p_a^{ss}, p_s \to p_s^{ss}\right\} \\
a_{22} &= \mathtt{D}\left[f^a\left(\bar{k}^w, p_a, p_s\right), p_a\right]/.\left\{\bar{k}^w \to \bar{k}^{w,ss}, p_a \to p_a^{ss}, p_s \to p_s^{ss}\right\} \\
a_{23} &= \mathtt{D}\left[f^a\left(\bar{k}^w, p_a, p_s\right), p_s\right]/.\left\{\bar{k}^w \to \bar{k}^{w,ss}, p_a \to p_a^{ss}, p_s \to p_s^{ss}\right\} \\
a_{31} &= \mathtt{D}\left[f^s\left(\bar{k}^w, p_a, p_s\right), \bar{k}^w\right]/.\left\{\bar{k}^w \to \bar{k}^{w,ss}, p_a \to p_a^{ss}, p_s \to p_s^{ss}\right\} \\
a_{32} &= \mathtt{D}\left[f^s\left(\bar{k}^w, p_a, p_s\right), p_a\right]/.\left\{\bar{k}^w \to \bar{k}^{w,ss}, p_a \to p_a^{ss}, p_s \to p_s^{ss}\right\} \\
a_{33} &= \mathtt{D}\left[f^s\left(\bar{k}^w, p_a, p_s\right), p_s\right]/.\left\{\bar{k}^w \to \bar{k}^{w,ss}, p_a \to p_a^{ss}, p_s \to p_s^{ss}\right\}
\end{aligned}$$

For the numerical example of Chapter 7, this system yields three eigenvalues, one of which is negative. Choose the slope of the eigenvector vector associated with the negative eigenvalue as in (9.15) and (9.16). Denote these slopes as slope$p_a \equiv v_{21}/v_{11}$ and slope$p_s \equiv v_{31}/v_{11}$. The syntax for integrating the two policy functions *backward* from the steady-state p_a^{ss} and p_s^{ss} to obtain the p_a, and p_s values in the interval $[k(0), k^{ss}]$ is the following:

$$\begin{aligned}
ss &= \texttt{NDSolve}[\{P^{a\prime}[\bar{k}^w]== \\
&\quad \texttt{If}\left[\bar{k}^w==\bar{k}^{w,ss}, \texttt{slope}p_a, \frac{f^a\left[\bar{k}^w, P^a[\bar{k}^w], P^s[\bar{k}^w]\right]}{f^k[\bar{k}^w, P^a[\bar{k}^w], P^s[\bar{k}^w]]}\right], \\
P^{s\prime}[\bar{k}^w] &= =\texttt{If}\left[\bar{k}^w==\bar{k}^{w,ss}, \texttt{slope}p_s, \frac{f^s\left[\bar{k}^w, P^a[\bar{k}^w], P^s[\bar{k}^w]\right]}{f^k[\bar{k}^w, P^a[\bar{k}^w], P^s[\bar{k}^w]]}\right], \\
P^a[\bar{k}^{w,ss}] &= =p_a^{ss}, P^s[\bar{k}^{w,ss}]==p_s^{ss}\}, \{P^a, P^s\}, \{\bar{k}^w, \bar{k}^w(0), \bar{k}^{w,ss}\}]
\end{aligned} \tag{9.18}$$

The solution is the sequence $\{p_a, p_s\}$ over the range $[k(0), k^{ss}]$. Knowing this sequence, recall the differential equation

$$\dot{\bar{k}}^w(t) = f^k\left(\bar{k}^w(t), p_a(t), p_s(t)\right)$$

and solve it for $\bar{k}^w(t)$, for all $t \in [0, T]$, $T < \infty$, *given* the solution (9.18) as follows:

$$\begin{aligned}
ss2 &= \texttt{NDSolve}[\{\bar{k}^{w\prime}[t]==f^k\left[\bar{k}^w[t], P^a[\bar{k}^w[t]]/.ss, P^s[\bar{k}^w[t]]/.ss\right], \\
\bar{k}^w[0] &= =\bar{k}^w(0)\}, \bar{k}^w, \{t, 0, T\}]
\end{aligned}$$

We now have a time series for the optimal solution $\left\{\bar{k}^w(t)\right\}_{t\in[0,T]}$ for large T ($T \to \infty$ at the limit) which is "stored" in the register $ss2$. We next construct the optimal sequence $\{p_a(t), p_s(t)\}_{t\in[0,\infty)}$ from

$$\begin{aligned}
p_a(t) &= P^a(k(t))_{t\in[0,\infty)} \\
p_s(t) &= P^s(k(t))_{t\in[0,\infty)}
\end{aligned}$$

using the syntax

$$\begin{aligned}
p_a[t_] &= (P^a[k[t]]/.ss2/.ss) \\
p_s[t_] &= (P^s[k[t]]/.ss2/.ss)
\end{aligned}$$

9.2 Backward integration method

One important shortcoming of the time elimination method described in Section 9.1 is that it is suitable only for solving monotonous adjustment systems with a single-dimensional stable manifold (Brunner and Strulik, 2002). If the system has non-monotonous (e.g. cyclical) adjustment with multiple state variables i.e., solution is found on a multi-dimensional stable manifold (rather a *surface* in this case), Brunner and Strulik argue that the policy function cannot be obtained by elimination of time from the system of differential equations. Not being able to solve non-autonomous systems (with time t as an independent variable) is another limitation of the time elimination method.[9] An alternative method that Brunner and Strulik suggest that overcomes these limitations is the "backward integration method", which involves tracking down the solution path for the dynamic optimization problem starting from a value in the $\varepsilon-$neighborhood of the steady state back to the initial value(s) of the state variable(s), k_0. Instead of time elimination, Brunner and Strulik employ "time reversal", i.e. reverse the flow of the differential equation system that characterizes the equilibria, so that standard forward integration can be used to find the solution trajectory back to the initial condition k_0 starting from a point close to the steady state. A second time reversal will yield the solution trajectory converted back into forward looking time.

Consider the effect of time reversal on our standard system (9.1) and (9.2) by substituting $-\tilde{t} = t$:

$$\begin{pmatrix} \dot{\tilde{k}} \\ \dot{\tilde{p}} \end{pmatrix} = -P(\tilde{k}, \tilde{p}) \qquad (9.19)$$

where $\tilde{k} = k(-\tilde{t})$ and $\tilde{p} = p(-\tilde{t})$. With the multiplication of P with -1, the stable arm that would contain the solution trajectory in the original problem is now the unstable arm. Hence, the

[9] Roe and Smith (2008) used this soluton procedure in their study of the effect of HIV/AIDs on the South African economy.

unstable arm in the transformed system is actually the stable arm in the original system. A point $(k(t), p(t))$ that we would expect to find on the solution trajectory of the original system would correspond to a solution point $(\tilde{k}(\tilde{t}), \tilde{p}(\tilde{t}))$ on the unstable arm of the transformed system in reversed direction.

We can then choose an initial value on the unstable arm

$$\begin{pmatrix} \tilde{k}(0) \\ \tilde{p}(0) \end{pmatrix} = \begin{pmatrix} \tilde{k}^0 \\ \tilde{p}^0 \end{pmatrix}$$

in an $\varepsilon-$neighborhood of the steady state. The solution of (9.19) follows the same path as the solution of the original problem in reversed direction and passes through initial condition k^0 for an endogenously determined $\tilde{t}_N > 0$. Here, the important challenge is to choose the initial value (i.e. choose the value of the control $\tilde{p}(0)$ corresponding to the state variable $\tilde{k}(0)$), since there is no explicit analytical form representing the unstable arm of the transformed system. A possible initial value which is close to the steady state and close to the unstable arm (called the *starting value* rather than an initial value in ibid.) is given by

$$\begin{pmatrix} \bar{k} \\ \bar{p} \end{pmatrix} = \begin{pmatrix} k^{ss} \\ p^{ss} \end{pmatrix} + \varepsilon \begin{pmatrix} \upsilon_{11} \\ \upsilon_{21} \end{pmatrix}$$

where $\varepsilon \in \mathcal{R}$, and $\binom{\upsilon_{11}}{\upsilon_{21}}$ is the eigenvector associated with the negative eigenvalue corresponding to the Jacobian matrix $DP(k^{ss}, p^{ss})$ at the steady state. Here, the sign of ε depends on the selection of the direction of the departure from the steady state towards the initial condition k^0.

Now that we have obtained a starting value $(\bar{k}, \bar{p})$, we could simply integrate $-P$ towards the stopping criterion k^0 for some endogenous stopping time $\tilde{t}_N$. The solution will be a triplet $(\tilde{k}^i, \tilde{p}^i, \tilde{t}^i)_{i=1}^n$, $n \in N$, where n is the length of the solution sequence. The last step involved in the solution algorithm is to perform a second time reversal so that the trajectory is transformed back to forward looking time:

$$\left.\begin{aligned} k^i &= \tilde{k}^{n-i+1} \\ p^i &= \tilde{p}^{n-i+1} \\ t^i &= \tilde{t}^n - \tilde{t}^{n-i+1} \end{aligned}\right\} \text{ for } i = 1, ..., n$$

This final step yields the solution to the original problem.

If the system involves multiple state variables and the stable manifold is a plane rather than a curve,[10] then generating the starting value requires a mapping that allows for a cyclical structure around the steady state (ibid.). For the case of a two-dimensional stable manifold, the starting value is given by

$$\begin{pmatrix} \bar{k}(\theta) \\ \bar{p}(\theta) \end{pmatrix} = \begin{pmatrix} k^{ss} \\ p^{ss} \end{pmatrix} + \varepsilon(\sin\theta)v^1 + \varepsilon(\cos\theta)v^2, \quad \varepsilon \in \mathcal{R}, \quad \theta \in [0, 2\pi)$$

Here, v^1 and v^2 represent the two eigenvectors associated with the two eigenvalues with negative real parts corresponding to the Jacobian of the system.

The Matlab code files for the models in Brunner and Strulik (ibid.) and other applications can be found at the web-address http://kaldor.vwl.uni-hannover.de/holger/software/index.php

---

[10] An alternative procedure for solving multidimensional, infinite-horizon optimal control problems, called the relaxation procedure, has been proposed by Trimborn et al. (2008).

References

Aghion, P., and P. Howitt. 1998. Endogenous Growth Theory. Cambridge, MA: MIT Press.

Aghion, P., and P. Howitt. 2009. The Economics of Growth. Cambridge, MA: MIT Press.

Atkeson, A., and P. Kehoe. 2000. Paths of Development for Early-and Late-Bloomers in a Dynamic Heckscher-Ohlin Model, Federal Reserve Bank of Minneapolis, Staff Report 256.

Aumann, R. J. 1985. What is Game Theory Trying to Accomplish?, in K. J. Arrow. and S. Honkapohja (Eds.). Frontiers of Economics. New York, NY: Basil Blackwell. pp. 28–88.

Badri, N. G., and T. L. Walmsley, Editors (2008). Global Trade Assistance, and Production: The GTAP 7 Data Base, Center for Global Trade Analysis, Purdue University, West Lafayette Indiana.

Barro, R. J. 1974. Are Government Bonds Net Wealth? Journal of Political Economy. Vol. 81. Dec. pp. 1095–1117.

Barro, R. J., and X. Sala-i-Martin. 2004. Economic Growth. 2nd Ed. Cambridge, MA: MIT Press.

Barro, R. J., N. G. Mankiw, and X. Sala-i-Martin. 1995. Capital Mobility in Neoclassical Models of Growth. *American Economic Review.* Vol. 85. pp. 103–115.

Bhagwati, J., A. Panagariya, and T.N. Srinivasan. 2004. The Muddles over Outsourcing. *Journal of Economic Perspectives.* Vol. 14. no. 4. pp. 93–114.

T.L. Roe et al., *Multisector Growth Models*, DOI 10.1007/978-0-387-77358-2,

Brunner, M., and H. Strulik. 2002. Solution of Perfect Foresight Saddlepoint Problems: A Simple Method and Applications. *Journal of Economic Dynamics and Control.* Vol. 26. pp. 737–753.

Caselli, F. 2005. Accounting for Cross-Country Income Differences. In P. Aghion and S. N. Durlauf (Eds.). Handbook of Economic Growth. Amsterdam: Elsevier North-Holland. pp. 679–738.

Caselli, F., and J. Ventura. 2000. A Representative Consumer Theory of Distribution. *American Economic Review.* Vol. 90. pp. 909–926.

Cass, D. 1965. Optimum Growth in an Aggregative Model of Capital Accumulation. *Review of Economic Studies.* Vol. 32. pp. 233–240.

Chambers, R. G. 1988. Applied Production Analysis: A dual approach. Cambridge, New York: Cambridge University Press.

Chipman, J. S. 2007. Protection and Exchange Rates in a Small Open Economy. *Review of Development Economics.* Vol. 11. no. 2. pp. 205–216.

Cornes, R. 1992. Duality and Modern Economics. Cambridge, New York: Cambridge University Press.

Romer, D. 2005. Advanced Macroeconomics. Boston, MA: McGraw Hill.

Diewert, W. E. 1973. Functional forms for Profit and Transformation Functions. *Journal of Economic Theory.* Vol. 6. pp. 284–316.

Diewert, W. E. 1974. Applications of Duality Theory, in M.D. Intriligator and D.A Kendrick (Eds). Frontiers of Quantitative Economics. Vol. 2. Amsterdam: North-Holland. pp. 106–171.

Echevarria, C. 1995. Agricultural Development vs. Industrialization: Effects of Trade. *Canadian Journal of Economics.* Vol. XXVIII, no. 3. pp. 631–647.

Echevarria, C. 1997. Changes in Sector Composition Associated with Economic Growth. *International Economic Review.* Vol. 38, no. 2. pp. 431–452.

Echevarria, C. 2000. Non-homothetic Preferences and Growth. *Journal of International Trade & Economic Development.* Vol. 9. no.2. pp. 151–171.

Fare, R., and S. Grosskop. 2004. New Direction: Efficiency and Productivity. Boston, MA: Kluwer-Academic Publishers.

Fei, J. C., and G. Ranis. 1961. A theory of Economic Development. *American Economic Review.* Vol. 51. pp. 533–565.

Gaitán, B., and T. L. Roe. 2005. Natural Resource Abundance and Economic Growth in a Two Country World. Bulletin 05-1. Economic Development Center, University of Minnesota.

Gaitán, B., and T. L. Roe. 2007. Path Independence Among Early and Later Bloomers in a Dynamic Heckscher-Ohlin Model. Bulletin 07-1. Economic Development Center, University of Minnesota.

Giovannini, A. 1985. Saving and the Real Interest Rate in LDCs. *Journal of Development Economics.* Vol. 37 pp. 63–87.

Gollin, D., S. L. Parente, and R. Roberson. 2004. Farm Work, Home Work and International Productivity Differences. *Review of Economic Dynamics and Control.* Vol. 9 pp. 827–850.

Hall, R., and C. Jones.1999. Why Do Some Countries Produce So Much More Output Per Worker Than Others?. *Quartery Journal of Economics.* Vol. 114. no. 1. pp.83–116.

Hansen, L. P., and J. J. Heckman. 1996. The Empirical Foundations of Calibration. *Journal of Economic Perspectives.* Vol. 10. no. 1. pp. 87–104.

Hirschman, A. O. 1958. The Strategy of Economic Development. New Haven, CT: Yale University Press.

Irz, X., and T. L. Roe. 2000. Can the World Feed Itself? Some Insights from Growth Theory. *Agrekon.* Vol. 39. no. 4. pp. 513–28.

Irz, X., and T. L. Roe. 2005. Seeds of growth? Agricultural Productivity and the Transitional Dynamics of the Ramsey Model. *European Review of Agricultural Economics.* Vol. 32 no. 2. pp. 143–165.

Jones, C. 2007. Input-Output Multipliers, General Purpose Technologies, and Economic Development. Working Paper. Department of Economics, U.C. Berkeley and NBER.

Jorgenson, D. W. 1967. Surplus Agricultural Labor and the Development of a Dual Economy. *Oxford Economic Papers.* Vol. 19. pp. 288–312.

Jorgenson, D. W. 2005. Accounting For Growth In The Information Age. In P. Aghion and S. N. Durlauf (Eds.). Handbook of Economic Growth. Vol. 1A. Amsterdam: North-Holland. pp. 743–815.

Judd, K. L. 1998. Numerical Methods in Economics. Cambridge, MA: MIT Press.

Jorgenson, D. W. 1967. The explanation of productivity change. *Review of Economic Studies.* Vol. 34. no. 99. pp. 249–280.

Kehoe, T. J. 1996. Social Accounting Matrices and Applied General Equilibrium Models. Federal Reserve Bank of Minneapolis, Working Paper 563.

Kehoe, T. J. 2003. An Evaluation of the Performance of Applied General Equilibrium Models on the Impact of NAFTA. Federal Reserve Bank of Minneapolis, Research Department Staff Report 320.

Kıpıcı, A. 1996. Terms of Trade and Economic Fluctuations. The Central Bank of the Republic of Turkey. Research Department. Discussion Paper No. 9615.

King, R. G., and S. Rebelo. 1993. Transitional Dynamics and the Economic Growth in the Neoclassical Model. *American Economic Review.* Vol .83. pp. 908–931.

Koopmans, T. C. 1965. On the Concept of Optimal Economic Growth. In The Econometric Approach to Development Planning. Amsterdam: North-Holland.

Kydland, F. E., and E. C. Prescott. 1982. Time to Build and Aggregate Fluctuations. *Econometrica.* Vol. 50. pp. 1345–1370.

Lewis, W. A. 1954. Economic Development with Unlimited Supplies of Labor. *Manchester School of Economics and Social Studies* Vol. 22. pp. 417–419.

Li, N., and T. L. Roe. 2006. Validating Dynamic General Equilibrium Model Forecasts. Economic Development Center Working Paper, Department of Economics and Department of Applied Economics, University of Minnesota.

Lin, L. 1989. A Concordance Correlation Coefficient to Evaluate Reproducibility. *Biometrics.* Vol. 45. pp. 255–268.

Mas-Colell, A., M. D. Whinston, and J. R. Green. 1995. Microeconomic Theory. New York, Oxford: Oxford University Press.

Matsuyama, K. 1992. Agricultural Productivity, Comparative Advantage, and Economic Growth. *Journal of Economic Theory.* Vol. 58. pp. 317–334.

Mulligan, C. B., and X. Sala-i-Martin. 1991. A note on the time elimination method for solving recursive dynamic economic models. NBER Technical working Paper No. 116.

Mulligan, C. B., and X. Sala-i-Martin. 1993. Transitional Dynamics in two sector models of endogenous growth. *Quarterly Journal of Economics.* Vol. 108. pp. 739–773.

Nelson, H., T. L. Roe, and X. Diao (2009). Economic Growth and the Distribution of Income: A Growth Model Fit to Ghanaian Data. Discussion paper, Development Strategy and Governance Division, International Food Policy Research Institute. Washington DC.

Nichols, D. A. 1970. Land and Economic Growth. *The American Economic Review,* Vol. 60. no. 3. pp. 332–340.

Ramsey, F. 1928. A Mathematical Theory of Saving. *Economic Journal.* Vol. 38. pp. 543–559.

Robinson, S. 1989. Chapter 18: Multisector Models, Part 3: Input–Output and Social Accounting. In H. Chenery and T. N. Sirinivasan (Eds.) Handbook of Development Economics. Vol. 2. Amsterdam: North Holland. pp. 885–939.

Roe, T. L., and R. B.W. Smith. 2008. Disease Dynamics and Economic Growth. *Journal of Policy Modeling.* Vol. 30. pp. 145–168.

Romer, D. 2006. Advanced Macroeconomics. 3nd ed. Boston, MA: McGraw-Hill.

Romer, P. 1990. Endogenous Technological Change. *Journal of Political Economy.* Vol. 98. pp. 71–102.

Saracoğlu, D. Ş. 2008. The Informal Sector and Tax on Employment: A dynamic general equilibrium investigation. *Journal of Economic Dynamics and Control.* Vol. 32. pp. 529–549.

Sargent, T. J. 1979. Macroeconomic Theory. New York, NY: Academic Press

Sydsæter, K., A. Strøm, and P. Berck. 1999. Economists' Mathematical Manual. Berline: Springer.

Tesfatsion, L., and K. Judd (eds). 2006. Handbook of Computational Economics. Vol. 2. Amsterdam: North-Holland.

Trimborn, T., K. J. Koch, and T. M. Steger. 2008. Multidimensional Transitional Dynamics: A Simple Numerical Procedure. *Macroeconomic Dynamics.* Vol. 12. pp. 301–319.

Usher, D. 1980. *The Measurement of Capital.* Usher (Ed.) National Bureau of Economic Research. Studies in Income and Wealth, Vol. 45. Chicago: University of Chicago Press.

Watson, M. W. 1993. Measures of Fit for Calibrated Models. *Journal of Political Economy.* Vol. 101. pp. 1011–1041.

Whalley, J. 1988. Lessons from General Equilibrium Models. in H.J. Arron, H. Galper, J. A. Pechmand (Eds.) Uneasy Compromises: Problems of A Hybrid Income-Consumption Tax. Washington, DC: Brooking Institute. pp. 15–20.

Woodland, A. D. 1982. International Trade and Resource Allocation. Amsterdam, New York: North-Holland.

Yang, J. 2008. An Analysis of So-Called Export-led Growth, IMF Working Paper WP/08/220.

Yi, Kei-Mu. 2003. Can Vertical Specialization Explain the Growth of World Trade? *Journal of Political Economy.* Vol. 3. no. 1. pp. 52–102.

Author Index

Subject Index

E

F

H

FSC
www.fsc.org
MIX
Papier | Fördert
gute Waldnutzung
FSC® C083411

Kolibri 360 GmbH
Ferdinand-Jühlke-Straße 7
99095 Erfurt, Deutschland
produktsicherheit@kolibri360.de